LOCAL AREA
NETWORKS

LOCAL AREA NETWORKS

Developing Your System for Business

DONNË FLORENCE

and

CAL INDUSTRIES

WILEY

A Wiley-Interscience Publication

JOHN WILEY & SONS

New York · Chichester · Brisbane · Toronto · Singapore

Library of Congress Cataloging in Publication Data:

Florence, Donnë.
    Local area networks.

    "A Wiley-Interscience Publication."
    1. Local area networks (Computer networks)
2. Business—Data processing.   I. Title.

TK5105.7.F56   1989   004.6'8   87-34535
ISBN 0-471-62466-7 (pbk)

Printed in the United States of America

10 9 8 7 6 5 4 3 2 1

Contents

SECTION 4: AFTER THE LAN, WHAT?

CHAPTER 10   LINKING LANs   123

CHAPTER 11   UNIVERSAL CONNECTIVITY   131

Preface

Did you hear the one about the CEO who said, "If they had told me twenty years ago I'd need a degree in engineering *and* an M.B.A. to succeed in business in the eighties—why, I'd have gone into politics."
Maybe that CEO is you.

■ **Good News.**   You *don't* need to be an engineer to identify the ways a local area network can improve your business. The first thing you need to do is what you probably do better than anyone else: List the separate tasks and operations the workers in your company perform.

■ **More Good News.**   You don't need to be an expert on data communications, either, to relate your list to the technological capabilities of a LAN. This book shows you how to itemize and evaluate the jobs within your company to single out those tasks that could be performed more efficiently with a local area network in place. Now you're positioned to calculate the payback period for your LAN investment.

As you can see already, instructions for performing these jobs are presented here in a language with which you're familiar: Businessman's English. You won't need to be fluent in telecommunications technobabble — it's not your job to speak glibly of "asynchronous throughput ratios" — to decide if a LAN is an appropriate solution for your business.

■ **The Best News.**   If a LAN *is right* for your company, you will have already done half the work of selecting one. This book will take you the rest of the way: explaining your options, showing you how to describe your needs to vendors, and helping you integrate the LAN you select into the day-to-day operations of your business.

How to Use This Book

This book is divided into four sections.

> * Section One: Who Needs a LAN?
>
> * Section Two: Which Kind of LAN?
>
> * Section Three: Whose LAN Is It Anyway?
>
> * Section Four: After the LAN, What?

■ *Section One.*   *Who Needs a LAN?* will help you ferret out the operations in your company that could be performed more efficiently with a LAN in place. This is a critical first step and one that's too often neglected when businessmen find themselves seduced by the "state-of-the-art" bafflegab of technology vendors and their own communications technicians.

Make notes as you read Section One, and as you talk it over with others in your company. You may be surprised at what you discover about how certain jobs get done. Even if a LAN is not the way to improve your procedures, you'll have a good fix on which ones need improving. If your answer to "Who Needs a LAN?" comes up "Not me," you can skip the rest of the book, but you won't be sorry you spent time reading Section One.

■ *Section Two.*   If your answer comes up, "Yes, my business could see improvements that would make a LAN a worthwhile investment," you'll turn to *Section Two: Which Kind of LAN?* If you've tried to read other books and magazine articles on local area networking, you probably think this is a subject that can't be explored in plain English. The remarkable news here is that there are really only four critical variables in LAN options: media, topology, access method, and modulation. Not only are these variables explainable, but, even more important, the choice in each of these four categories is dictated by existing business factors, not technological hocus-pocus. You'll use those same function lists you created in Section One—and one or two other lists of facts about your company—to decide which kind of LAN is appropriate.

Truth is, when these four variables are explored in terms of "What do you want the LAN to do?"—the appropriate decisions become obvious. You'll look like a wizard for making the right choice, and no one has to know you don't know a Network Interface Unit from a curve ball.

■ **Section Three.** Another nontechnical subject you already know about comes into play—or should—when your company implements a networking system. *Section Three: Whose LAN Is It Anyway?* explores the organizational politics of sharing facilities and information via a local area network. The technophobia of some staffers, pitted against the technical "witch doctor" status of others, can spell disaster (and a substantial LAN investment down the tubes) if potential political problems aren't addressed during network selection and implementation. You'll apply your own good corporate-politics sense (you may want to shred the lists you make in Section Three) to the unique turf battles that can surface when a local area network is introduced.

■ **Section Four.** *After the LAN, What?* looks at your network investment as a building block in a larger communications web. Just as your decision to install a LAN is spurred by a perceived need to connect previously separate information centers, you will want, perhaps soon, to link those units with others—to make them part of a network that is not "local," but regional, national or even global in scope. You'll be making more lists in Section Four, naturally.

Expanding your network needn't mean ripping out the LAN hardware and software you've purchased or leased. The right planning can make your LAN just the first phase in an *incremental capital-investment strategy* whose return will make you look like a genius even if you never find out what a multi-channel, unidirectional, RF-analog, half-duplex signal is. Who wants to spout phrases like that, anyway, when "incremental capital-investment strategy" rolls so easily off the tongue?

Sections One through Four are not going to make you an overnight telecommunications expert, but you don't want to be. What this book aims to do is demystify the language of LANs, so that when you talk (in English) about your company's needs, you won't be left in the dust by answers that your ear hears as utterly meaningless phrases—"eye Tripoli ate oh to dot for." You will know as much as you need to know to match what the technology offers to what your business needs.

Some of you will find, after you've read Section Four, that you do want to know more about the technology of LANs—not because you

need to know more to make an intelligent purchase, but because
...well, it's what mountain climbers say about Mount Everest. For
you, some additional resources are listed at the back of this volume.

There's also, at the back of this book, a list of selected LAN vendors and their offerings. This implies no endorsement of any products or services available from these suppliers. Rather, the list is a
starting point, to help you get up-to-the-minute information about
the capabilities of specific networks. If you've read this book (and
made your lists) before you contact suppliers, you'll be prepared to
ask the appropriate questions and get meaningful answers. Then
you can select and implement the network that will enhance your
business and—we're talking a *substantial delta* here—your reputation
as a forward-thinking executive.

The Design of the Book

This book is designed for you to use as a reference, as a source of
information, and as a practical way to move through the process
of exploring the question, "is a LAN right for us" and if so "which
one?" Many of the pages within contain checklists and forms for
your duplication. With a multiple number of application possibilities, you will need a process and support tools to guide your company through the LAN exploration – and this book has been
designed to practically assist you in that process.

Donnë Florence

Acknowledgments

Special thanks are due to Michael Cotter and Roberta A. Vogel of NYNEX, Roger Freeman of Raytheon, and Kenneth A. Smalheiser of World Communication Works for their advice and encouragement. Researcher Theresa Conlon and editors Stephen Lyon Crohn and Janet Witter also provided especially valuable assistance in the preparation of this manuscript.

LOCAL AREA NETWORKS

SECTION ONE:
WHO NEEDS A LAN?

Chapter 1

What a Local Area Network Is...
and Isn't

Advertisements for local area networks used to appear only in specialized business magazines, where they aimed to catch the eye of information managers and data processing chiefs. These days, the ads appear in major metropolitan newspapers and general business magazines and the sales prospect is you (Fig. 1-1).

"Why me?" you wonder as you scan a headline ballyhooing the advantages of in-place wiring for your LAN. Here's a vendor pitching you a padded leather interior when you haven't even decided yet if you want or need a car.

Maybe you think these ads are aimed at executives who have already decided they want to install a local area network. Yes, of course, there are business leaders whose own technical staffs have convinced them the company needs a LAN, but those executives are a small segment of the audience that will see these ads. The larger segment—the primary audience for this advertising—is you. In that case, why don't these advertisers begin their campaigns by explaining what a local area network is and why your business might want one? It's more cost-effective *not* to, that's why.

For example, if Vendor A is using its expensive full-page space in *The New York Times* to explain the ins and outs of LANs, there's no space left to explain what differentiates Vendor A's product from Vendor B's. Now that you've read Vendor A's ad and you understand LANs, you turn to Vendor B's product-differentiation message, and you're sold—on Vendor B's LAN. Nobody in the industry wants to be Vendor A.

Besides, if you really want to know whether a LAN could be a useful tool for your business, you'll make it your business to find out: You'll ask your telecommunications manager, he'll make a speech you won't understand about NIUs and transceivers, you'll give him a raise to hang onto his expertise and delegate the LAN-selection job to him. See? There's no need for the advertisers to tell you what a LAN is or why you might want one.

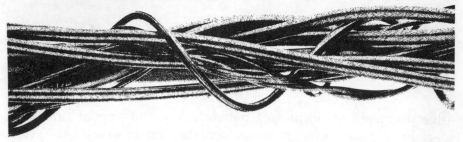

Fig. 1-1

Even if you don't want to hear the speech or give your telecom manager a raise, there's still a bandwagon effect to this type of advertising. By not explaining what a LAN can do for your business, advertisers imply that you already know. They know you want to believe this: You're a savvy, leading-edge executive, now you can proceed to Step Two, product selection. Go ahead, cast your vote for the most effective advertising; maybe no one will ever figure out that you bought a Volkswagen when you needed an 18-wheeler, or that you paid for a Rolls-Royce when a bicycle would have been adequate.

That doesn't mean LAN vendors are a duplicitous bunch who'll do or say anything for a sale. Chances are you use similar techniques in advertising your own products and services—or would if you could. But here's where it leaves you: You don't want to commit corporate resources to another technological breakthrough without knowing just what that breakthough can do for your company's bottom line. You need some answers before you study those product-differentiation ads.

A WAY TO COMMUNICATE

First, you need to know that a local area network is a way to let nearly every data processing and data communications device in your organization exchange information with the other devices on the network.

You probably already have wires of some kind running from one device to another—from a personal computer to a printer, for example. Now imagine what a jumble of wires you'd have if you ran separate cables from each machine to all the other machines (Fig. 1-2).

If the picture in your mind right now is an unholy mess, think about this: Suppose you could build a superhighway for this job —a single express road all your equipment could share to communicate even more efficiently than with dozens or hundreds of separate "point-to-point" connections. The express road you're visualizing is a local area network (Fig. 1 3).

CRITICAL CHARACTERISTICS

Next, you need to know the seven characteristics that all local area networks have in common.

1. **Local** – up to 2-mile range

2. **Private** – user-owned and -controlled

3. **Structured** – pathway that connects independent devices

4. **High-speed** – typically 1, 4, 10, or 100 Mbps

5. **Commercially available** – can be purchased

6. **Transmit packets of data** – for most efficient use of the network. Even networks that are not truly "packet switched" are programmed to operate as if they were.

7. **Connectable** – to link one device to others

Fig. 1-2. No.

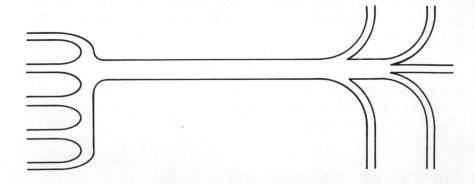

Fig. 1-3. Yes.

1. Local

For starters, a local area network is **local**. (This is going to be easier than you thought, isn't it?) That means it links data equipment (computers, terminals, printers and so on) that's fairly close together, but far enough apart that the devices can't be easily linked in some other way. You would not install a LAN just to connect your home computer to your home printer a few feet away. An inexpensive printer cable is all you need for a simple point-to-point connection like this, even if your computer and printer are from two different manufacturers. Linking one computer to other input devices (a scanner or barcode reader, for example) or output devices (a graphics plotter or facsimile machine) probably won't require a local area network, either. Simpler, cheaper options are available to make these connections. Most networks are installed to link several computers with one another as well as with those peripheral devices.

A local area network may connect devices that are all in the same building, in separate buildings of a campus-like location, or in separate offices in the same town. Generally, the scope of a LAN won't exceed ½ mile, but there are ways to reach beyond that limit if needed.

2. Private

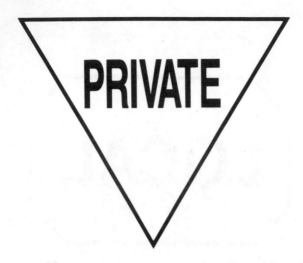

Second, a local area network is **private**. Unlike the public telephone network, which you may be using now to transmit some kinds of data, you don't share this highway with other user organizations. It's your company's private road. Your company owns its LAN, even if you've only leased the hardware and software that make up the network; your company administers its own LAN; and it will generally be limited to intracompany traffic. Just as two country neighbors might decide to build and share a private road between their properties and the main highway, so two or more companies, say, in the same high-rise office complex, can share the capital investment in a single LAN. Such a shared LAN might be set up so that their respective data would never be traveling on the same stretch of road at the same time. Thus, for each partner in such an enterprise, "its" LAN is privately owned, privately administered, and limited to intracompany traffic.

One other thing you should know about the private nature of a local area network: It is not subject to Federal Communications Commission or Public Utilities Commission regulation. This means you can't legally sell or lease time on your LAN to another customer. (An hour ago, you didn't know what a LAN was, now you want to compete with the local telephone company?) It also means that some of the non-regulated companies born of the AT&T divestiture are now in the business of selling, but not manufacturing, local area networks. You can expect to hear from them.

3. Structured

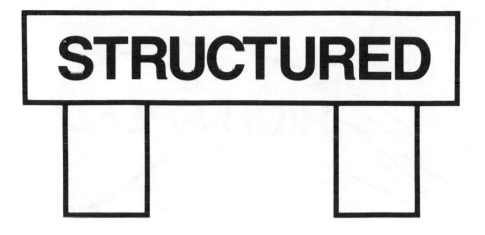

The third characteristic all LANs have is that they are **structured**. That is, a LAN is a discrete physical pathway. The devices it connects had "a life of their own" before the LAN was installed; they do not lose their capability to do the things they used to do just because a LAN is installed. A log cabin in the woods, is a log cabin in the woods, whether or not there's a road leading to it; it remains a log cabin in the woods after the road is built.

This may make you think that a LAN is a piece of hardware, a length of wire and some boxes you can hold in your hands. That is not entirely true. The "traffic cops" that keep data moving on this private road are actually instructions programmed into the network software. These instructions are an integral part of the "discrete physical pathway" even though you will probably never see them—and you certainly won't touch them. If you think about it, you never see or touch the electrical and electronic devices that control the stoplights in your city, either, but they are a vital part of the overall traffic system, just as network software is: They keep traffic moving efficiently where they are installed, and they aren't needed at all where there are no roads.

4. High-speed

Fourth, all LANs operate at **high speeds**. The speed at which data travels along a LAN or any other pathway is given in *bits per second*. The numbers that come up in discussions of network speed are staggering. They are also unintelligible if you can't relate them to something in real life. Try this: If one full printed page of this book has approximately 500 words, an average of 6 characters per word, and 8 bits per character, how long would it take to transmit the page at 2.4 kilobits (2400 bits) per second? Answer: 10 seconds. Is that fast or slow? Compared to what? Certainly it's faster than you could copy the same words, one keystroke at a time, into your computer. If the words are already stored on my computer, I can transmit them directly to your computer in 10 seconds, a substantial saving over the time it would take even the most accomplished keyboard wiz to type them.

Not fast enough? with a higher-speed link, say, 9600 bits per second (9.6 Kbps), I can transmit the same data to you in under 3 seconds. That's starting to be impressive until you think about how much—or in this case, how little—data we're sending along this pathway. One page? Suppose you need the whole book? Suppose you need your company's monthly regional sales reports? Yes, you can still get that data faster via an electronic link than via the interoffice mail, but is it fast enough?

4. High-speed (continued)

Most LANs operate at speeds of 1–10 megabits per second (Mbps). In computerspeak, mega is one million, kilo one thousand. At 1 Mbps, you could transmit all the words in this book in under 5 seconds. At 10 Mbps, in less than a second. If you can read the book that fast, you're not paying attention.

Do you need to transmit data that fast? It depends on what you're doing with the data, how many people in your organization need access to it, even what your competition is doing. These are all factors we'll explore in later chapters. For now, you just want to get a handle on the volume of data you can move rapidly—almost instantly—from one computer to another with a LAN in place.

Although 1–10 Mbps is the typical LAN speed range, some LANs operate at speeds as slow as 24 bps. If that's sufficient for your needs, you won't want to spring for a fiberoptic beauty that runs at 140 Mbps. Even 140 Mbps is not the upper limit in LAN speeds; developers have pushed experimental LANs to speeds measured in *giga*bits (billions of bits) per second.

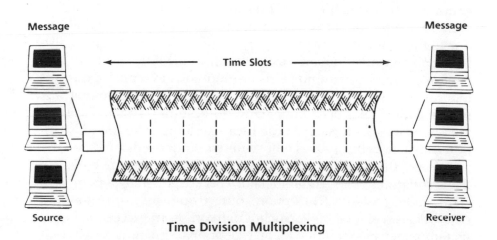

Time Division Multiplexing

5. Commercially available

Fifth (here's another easy one): For our purposes, all LANs are **commercially available**. If you decide a LAN would be useful to your company, you don't really want to wait two years for the vendor to finish inventing it. This part of the definition of LANs also excludes networks that are custom-built to meet the needs of one user organization, with no plans or intentions to replicate the system for another user organization. You should be aware that this can be done—it has been done—and the resulting system is a local area network. Such catch-as-catch-can systems were the only option available to businesses whose networking needs predated the creation of the commercial LANs we're talking about here. For today's computer users, commercial LANs are generally a more efficient, more cost-effective option, but customized, one-of-a-kind networks can still be built. Such a network is an exceptional (and usually expensive) solution for an exceptional situation, and its cost must be weighed against the feasibility of replacing the existing, oddball, hardware with off-the-shelf equipment that can be linked via a commercially available local area network.

6. Transmit packets of data

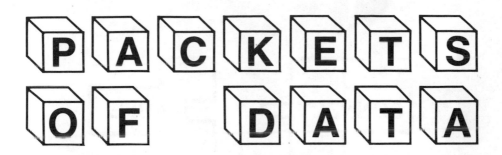

Sixth, local area networks **transmit packets of data**. They break a long transmission into smaller parts and reassemble them at the receiving end. This allows more efficient use of the network.

The most sophisticated networks are "packet-switched." Truth is, you could buy a LAN without knowing what this means, but somewhere in the selection process somebody will try to impress you with this phrase. The drop-dead comeback is, "How is that different from a circuit-switched network?" Just in case your vis-a-vis doesn't know, or tries to bluff you, here's the answer:

Your telephone system is *circuit-switched*. When you make or receive a phone call, "possession" of the line (circuit) is temporarily transferred to you exclusively. Once you have "seized the circuit," as the telecom people say, no conversation you're not part of can be transmitted along that pathway. When you hang up, the system can transfer the use of the circuit to someone else who needs it, but as long as you have your phone off the hook, that circuit is your private road. In effect, then, your telephone system is a lot of private roads, each carrying no more than one vehicle at any moment, and there's a lot of unused capacity on those roads.

Obviously, it would be more efficient if several vehicles could use one road simultaneously, provided they could stay out of one another's way. That's the object of "packetizing" data. The network software slices your data transmission into packets of bits, eases the packets in sequence into other traffic on the network, and reassembles them into a single transmission at the receiving end. Because LANs operate at such high speeds, neither you nor the receiving station will ever perceive that your transmission was not one continuous flow of data.

7. Connectable

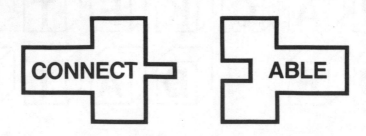

Finally, the trickiest one: A local area network **can connect every device** *on the network* **to every other device** *on the network*. The tricky part is that you can't necessarily connect *every device you own* to your LAN—at least, not yet. "Universal connectivity" is the goal toward which the entire telecommunications industry is moving, but that journey isn't over yet.

A LAN comprises not only the network software and the physical pathway from device to device, but also any intermediate boxes and cables that allow one device to communicate with the network. That is, it's not just the stoplights and the freeway, but also the surface streets, access roads, and on- and off-ramps that allow you to use the freeway. If your vehicle can't travel on those surface streets, it can't use the freeway.

Similarly, if your computer can't "speak the same language" (use the same protocols) as the network, it cannot be connected to the network without a special on-ramp called an *interface*. As LAN manufacturers move toward universal connectivity, more and more interfaces are available that allow more and more different brands of equipment to be connected to a single network, and innovative software engineers are creating new solutions to these problems —literally, every day. For now, however, you should know that you won't *necessarily* be able to link a hodgepodge of different vendors' equipment to a LAN. This may be an important consideration in your LAN-selection process.

WHAT A LAN CAN DO

Now that you have some idea of what a LAN—any LAN—is, let's look at what it can—and can't (or shouldn't)—do.

Data Transmission

A LAN is a pathway for the transmission of data. You know that data includes both words and numbers—the information that comes out of your company's computers. In telecommunications, "data" is *any information*, including sound and pictures, *that has been digitized*.

Your company may already be using digital telephone equipment—a system that converts the analog sound of your phone conversations to a digital signal that can be transmitted as a series of on/off bits. Think of the analog sound of your voice as a sine wave:

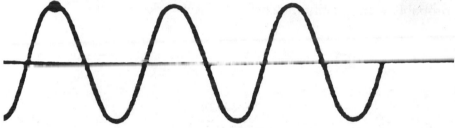

Fig. 1-4. Analog transmission.

Now lay a piece of graph paper over that sine wave, and convert it to a bar chart:

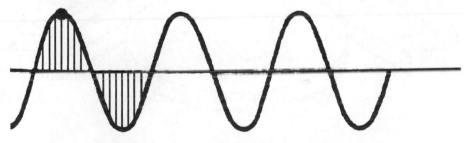

Fig. 1-5. Remodulated transmission.

That's a visual approximation of what happens when an analog signal is remodulated into a digital signal. (The process can be reversed, converting digital signals to analog.) Since voice signals can be digitized, you can use your LAN for voice communication, right? Yes, you can, but why *would* you when you already have a telephone system for that?

Sound Transmission

You may still want to use your LAN to transmit sound, including voice, other than garden-variety interoffice telephone calls. A LAN can encompass, for example, a security system that "listens" for intrusions or machine shutdowns. With the appropriate speakers and other hardware, it can include company-wide broadcasting or alarm systems, if you don't already have other facilities that serve the purpose.

Sound transmission should be a primary purpose of your local area network. The real forte of these networks is data transmission. Security listening systems and broadcast capabilities can be built into many of today's voice telephone systems, and stand-alone systems for these functions are also available. If you don't really need the data transmission capabilities of a LAN, this extra function probably won't justify the cost of the network.

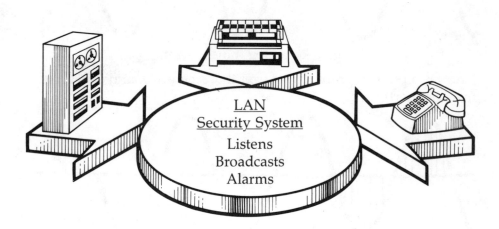

Video Transmission

The digital **video** capabilities of local area networks strike some executives as just plain silly. "Why would I want to buy a LAN to videoconference with my own staff," asks one, "when I can walk down the hall for free and talk to them in person?" Indeed, video is not for everyone. On the other hand, if video allows you to improve plant security or to conduct cost-effective product demonstrations for more employees, and if the incremental cost of adding this capability to your LAN is not prohibitive, it may be just the ticket for your company.

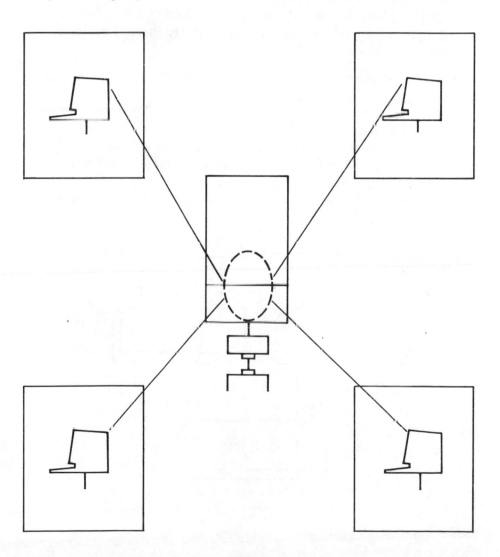

Shared Resources

Far and away, though, the most valuable function of a LAN is allowing users to **share electronic resources** efficiently.

- This means that every computer on your network can produce good-looking final copy on the one expensive laser printer you own. High-priced computer hardware is not the only "electronic resource" users need to share.

- It means every device on the network can share applications software. Your accounting department has fine-tuned a spreadsheet package to meet the unique requirements of your company; now everyone can use that software instead of the off-the-shelf version.

- Perhaps most important, it means that every computer on your network can share information. Everybody works from one up-to-the-minute set of sales figures to develop projections you know you can rely on. It's not easy to assign a dollar value to all the benefits of sharing information. How do you appraise the worth of time not lost, mistakes not made? One way, of course, is to analyze the cost of time that was lost, mistakes that were made. In Chapter 2, we'll look at some less painful tactics.

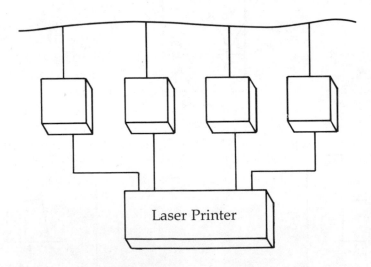

Laser Printer

CASE STUDY—Part I

Electromedia Inc. is a corporation that hitched its wagon to technology's star 30 years ago, publishing manuals and reference books for electronics buffs. Over the years it has acquired other, related businesses, including other electronics-book publishers, several electronics magazines, a couple of software companies, a direct-mail subsidiary, and a typesetting and printing company. Not all the acquired enterprises have been moved yet to corporate headquarters in Nyack, New York. The typesetting operation remains in West Nyack, about two miles from the main office; the software companies have been merged into one operation, located upstate in Rochester.

However illogical this may seem, given the nature of its business, Electromedia Inc. never computerized its operations in a unified, centralized way. The accounting department got its first minicomputer in 1960 and traded it in years ago for a newer, faster, more powerful model. Other departments and divisions computerized piecemeal. The circulation department has one terminal linked to the computers of an outside fulfillment service, but still ships printouts of magazine mailing labels to printers throughout the United States. The book companies and individual magazines have a variety of personal computers, electronic typewriters, and word processors, used with a wide range of skill by different staffers. (A couple of staff writers still turn in hand-written copy to secretaries who type it for the editors.) The typesetting center has sophisticated equipment for both word processing and typesetting, but only one of the magazines delivers "captured keystrokes" on floppy disks. All the other work delivered to the typesetting center must be re-keystroked by typesetting staffers. None of Electromedia's departments or divisions is electronically linked to any other, so all interdepartmental communication is handled by phone, face-to-face, or on paper. Voluminous printouts are shipped regularly between Rochester and Nyack, and Electromedia's President and CEO, Sylvester Fachs, still doesn't feel he can keep close enough tabs on the software operation.

This has the CEO wondering if all his company's electronic hardware couldn't be tied together in some useful, not too expensive, way that would allow more efficient use of those resources, better supervision of all the departments, less waste, and maybe even new profit opportunities. Perhaps a LAN is an appropriate solution.

End of case study (Part I)

Chapter 2

Analyzing Your Company's Operations

To determine whether a local area network is an appropriate tool for your business—and if so, where it can help most—you need to study your company's internal operation. You need to know:

- Who communicates with whom?

- Why?

- By what methods?

- How often?

- For how long?

- What is the dollar value of the time spent on each of those communications?

- Which of those communications could be speeded up with a LAN?

- What is the dollar value of the time that could be saved with a LAN?

Employee time saved is not the only measure of a network's value, but it is easier to calculate than the cost of mistakes not made or bad judgments not acted upon. The bottom line in this exercise allows you to calculate the approximate payback period for your LAN investment.

If the annual savings you could realize with a LAN are small, it could take your company years to recoup the cost of an extensive network. Even after you add in your best-educated guess on potential "soft" benefits—the opportunity to bring a new product to market faster than your competitors, the strategic advantage of better inventory control— the payback may stretch beyond your magic number of, say, five years, and you may decide a local area network is not right for your company.

Not so fast. In fact, a smaller network, for only part of your organization, might still produce the extra efficiency you want within a manageable payback period. If you've investigated the worth of installing a company-wide network, you will have already done the detective work to identify which part of your organization can make the most profitable use of a LAN.

WHAT KINDS OF COMMUNICATION HAPPEN?

Understand, unless yours is an extremely small organization, detective work is not too strong a phrase for what's required to answer the questions outlined above. You may think you know what each employee in your organization does all day long, but *do* you?

Consider your own work first. A simple way to track your communication with various people within your organization is to set up a chart as seen on pages 23 and 24.

Instructions on Filling Out Charts

1. In the first column list every person in your organization with whom you communicate—or try to—in the course of a single day.

2. What was the purpose of the message or conversation?

3. How did it occur (written, face-to-face, travel, phone, etc.)?

4. How much of your time did it take?

5. At the end of the day, consider how often you have communications just like each one you've listed.

6. How many hours a year do you spend sending and receiving information within your company?

Could a LAN save *you* time? How much time have you spent contacting several individuals with the same message? How much time have you spent preparing memos and getting them distributed? How much time have you spent just requesting and waiting for reports or other written materials you need to read? Once you got those materials, how much time did you spend preparing them so you could use or work on them (e.g., copying, keystroking information into a different computer, color coding or highlighting sections)? If you're using a personal computer, how much time have you spent swapping floppy disks with someone else? Given your compensation package (yes, including perks), how much does that time cost your company?

This is the kind of analysis you need to undertake for *all* your employees. If you really can write down the right answers without leaving your chair, your organization is probably too small to need a local area network. If your organization is a large one, this probably sounds like a year's work. Don't worry, you're going to have help, as you'll see in Chapter 3.

Brenda Santana — Office Manager
Your name and title

#1 Whom	#2 Purpose
Stacie Baron Prince Production Manager	• Phone Messages • Invoices • Packages
Janet Witter Editor in Chief	• Phone Messages • Correspondence • Information Gathering
Richard Lombard Sales Manager	• Phone Messages • Freight Reports • Invoices • Correspondence

#3 How	#4 How long	#5 How often
Phone/written	10 minutes	Every day
Phone/face to face	30 minutes	3 times a week
Face-to-face	5 minutes	Every day
Phone/written	10 minutes	Every day
Written	20 minutes	Every day
Face-to-face/written	40 minutes	Once a week
Phone/written	15 minutes	Every day
Written	25 minutes	Once a month
Face-to-face	30 minutes	3 times a week
Face-to-face/written	20 minutes	Every day

#6
Annual total

504

WHAT KINDS OF COMMUNICATION DON'T HAPPEN?

Day-to-day communication is not the only aspect of your operation that ought to be analyzed if you're thinking about a network. This is a bit trickier, but you ought to look at non-communication, too. Don't try to do this at the end of a long day, when your're tired, because you need to use your imagination. If you always get your most creative ideas in the shower, that's the time you want to tackle this job: Turn on a tape recorder (outside the shower) and do your thinking out loud. Ask yourself these questions:

- What information about my company could I use that I don't get at all? What could I do with that information?

- What information do I get at regular intervals that I could use more often? What difference would it make in my effectiveness?

- What information do I get that would be more useful if it were more timely, even if I didn't get it more often? How would that improve my company's profits?

- What information do people in my company have that I haven't asked them to share because there's no procedure for sharing it? Who else could do their jobs more effectively if this information could be shared?

- Does anyone else in my company have information about what our competitors are doing that isn't regularly shared? Is it information we could use?

- How much of this missing information already exists on data files somewhere in my company?

- How much of this missing information could exist on data files if we weren't wasting so much time carrying printouts and floppy disks around the office?

- How many times has our planning gone awry because two or more employees were working from two different sets of starting figures? How much time have we lost because we had to back up and start over?

These are all problems a local area network could help solve, and the solutions add value to your network. If you don't have these problems, then this added-value component won't figure in your LAN-payback calculations.

Turn to pages 26-30 for forms to copy and fill out regarding the above questions.

Survey your employees again. You may not be the only person in your organization with the creative imagination to ponder these non-communication questions. You may discover one of your secretaries has enhanced her word-processing software and that others are eager to get their hands on that application disk.

More investigative work, more time and effort to be spent on just figuring out if a LAN is a worthwhile investment for your company. Wouldn't it be cheaper just to buy a LAN than to invest this much time in deciding whether you ought to?

WHAT ABOUT INFORMATION I DO NOT RECEIVE?

Type of Information Not Being Received	What I Could Do with This	Value to Me and the Company

WHAT INFORMATION DO I GET AT REGULAR INTERVALS THAT I COULD USE MORE OFTEN?

Information	Current Frequency	Requested Frequency	Use and Value to Me and the Company

WHAT INFORMATION DO I GET THAT WOULD BE MORE USEFUL IF IT WERE MORE TIMELY?

Information	Time Being Recorded	Requested Time Pattern	Company's Profit with New Time Pattern

SHARED AND NOT SHARED

Unshared Information	Benefits of Sharing That Information	Information about Competitors That Isn't Normally Shared	Benefits of This Information to You and the Company

WHAT AMOUNT OF MISSING INFORMATION EXISTS?

Data Files in the Company	Time Wasted Carrying Printout and Floppy Disks	Amount of Lost Files	Time Lost: Employees Working from Different Starting Points	Benefit of Hooking Up a LAN

Maybe it would. You could buy a small network, install it in part of your organization and measure the results. Maybe you won't choose one that's too complex for the employees who need to use it. Maybe you won't install it in the one segment of your organization where the payback will be uncharacterisically high or low. Maybe an experiment like that will give you accurate, reproducible results that point the way unequivocally to the right company-wide decision. Then again, maybe it won't.

In fact, all this detective work needn't be a burden. If you decide to invest in one or more LANs, getting the network that serves you best will make this effort well worthwhile. In Chapter 3, we'll look at ways to distribute this labor and other benefits of involving a committee of users in this project.

CASE STUDY—Part II

Thinking about his own communications with others in his company, Sly Fachs doesn't see many opportunities for savings. He couldn't communicate electronically with his managers because his own personal computer is still in a box under his desk; he doesn't have time to read the user manual and learn how to use it. Thinking about other parts of his company's operation is more fruitful.

If the managers submitted their financial data electronically to the accounting department's mini, the computer could flag numbers that were out of line with budgets and corrective action could be initiated sooner.

If the accounting department could notify the magazine publishers electronically that certain accounts were in arrears, the publishers could put pressure on those clients to pay up—instead of continuing to accept additional advertising-space reservations from on-paying accounts.

If the editorial departments could transmit copy electronically to the typesetting center, the center could save the cost of rekeystroking (retyping) all that text (perhaps four of five full-time salaries). Maybe they wouldn't need a van and driver to shuttle back and forth all day between the two locations.

If it didn't take so long to send and receive information about the software subsidiary, maybe that operation could be turned into a profitable one.

Maybe the magazine mailing labels could be transmitted electronically to the printing houses.

Those weren't the only ideas that occurred to Sly in the shower one morning, but that was as many as he could remember and write down by the time he'd toweled off. And, he was sure, there were other inefficiencies in the company's operation that he didn't even know about, but his managers would. It was time to call them into a meeting to discuss the ways that a local area network—or some other electronic link—could improve the company's profitability.

End of case study (Part II)

Chapter 3

Organizing a User Committee for LAN Selection

If the word "committee" makes you cringe, you're not alone. Too often a committee is, as Russell Baker has put it, "a mutual protection society formed to guarantee that no one person can be held to blame for a botched job that one man could have performed satisfactorily." It needn't be so, however.

If you think about it, your company is a sort of committee, organized for the purpose of delivering whatever goods or services you provide. What makes this committee effective is that each member (employee) has specific responsibilities. You're not paying full-time salaries to people who do nothing more than sit around a table philosophizing once a month.

What makes most committees so ineffective is that each member believes philosophy is what he's expected to contribute. "Well," the new appointee says, "I really don't have time to *do* anything, but I'll be glad to *share my ideas* with the other committee members."

You'll undoubtedly hear this response when you begin to organize your LAN committee. You might be able to head that off by calling the group something else — LAN Feasibility Study Project or LAN Needs Assessment Work Force.

The critical fact about this committee is that each member has a specific job to do. No philosophers, no gray eminences, no designated "idea persons." The overall mission of the group is to identify and prioritize the operations in your company that could be most improved with the installation of a LAN.

NEEDS ANALYSIS

The committee's work begins with information gathering. Each member will interview a number of prospective network users how their communications time is allocated, as you interviewed yourself in Chapter 2. "Prospective network users" means employees who already have computers, terminals, word processors, or other data equipment in their offices. For this job, you'll want to reproduce as many copies as necessary of the same form you used.

Employee name and title

Whom?	Purpose?	How?	How long?	How often?	Annual total
1.	2.	3.	4.	5.	6.

This job does not require any technological sophistication on the part of committee members or the people they're interviewing, as you've already demonstrated. In fact, there's a big advantage to involving nontechnicians in this information-gathering phase of the project: They will use language the interviewees understand to phrase their questions, and they will relay nontechnical answers without trying to 'improve' the language. It's okay to report that Maria wants to have the latest version of a contract proposal on her computer without having to type in new versions before she can work on it. It's okay to say Joe wishes his computer could talk to Harry's; it isn't necessary to upgrade Joe's language to "an expressed desire for complementary data transfer."

That doesn't mean technicians can't or shouldn't contribute to the committee's work, even at this information-gathering stage. In fact, they ought to be involved at this stage if they're going to be involved later on. It will help them ease into an English-speaking environment. But don't send a systems analyst to interview clerical employees about their communications needs. The clericals will be intimidated and the information you get back will not be as accurate as it might appear to be. Let the technical members of your committee gather information from other technicians, who aren't on the committee.

It's likely that your company has employees whose jobs are *exactly* like someone else's. It may be unnecessary to interview all twenty quality assurance specialists, but it isn't wasteful to interview more than one of them. Some of your employees have more insight than others or may think more deeply about the nature of their work. What's more, you don't want to overlook the significance of having the same communications need identified by a large number of employees.

AN INTERIM REVIEW

After each committee member has surveyed his or her 'constituents' in this way, you'll want to look over the results and call the committee together for a brief meeting. The point of this meeting is not to exchange ideas, but to present a summary of findings to date about what kinds of communication take place in your company and how much time is spent on them.

The hidden agenda is to allow technical and nontechnical members to learn to speak to and understand one another. Toward that end, you'll call on specific committee members to comment on one or another of their findings.

"Marie, you wrote here that Joe wishes his computer could talk to Harry's. Do you know why Joe wants that?"

"Melvin, you've written that you need a LAN that will pave the way to interoperability and ISDN connection. Could you explain that in language we can all understand? Or is it something we don't need to consider yet?"

At the end of this meeting, you'll summarize:

"It appears that sixty percent of our clericals are spending up to an hour a day carrying floppy disks from one machine to another. All our managers are soliciting, and waiting for, reports from the data processing group. Nearly everyone in the company is playing telephone tag with our own sales force to get information the salespeople have stored in their portable computers.... These appear to be areas where a LAN could help us. Now we need to know more."

ADDITIONAL SURVEYS

The committee members' next assignment is to survey constituents with the "non-communication" questions you examined in Chapter 2. They'll be soliciting suggestions—even far-fetched ones—that will expose gaps in your company's current communications. You'll use that information in another summarizing meeting to outline additional ways that a LAN could make your operation more efficient, more profitable, or maybe just less aggravating.

By now you should have a fairly complete list of potential applications for a LAN in your company. Your list might include such items as:

- Shared access to accurate, up-to-date information in consolidated databases.

- Easier, wider access to scarce electronic resources, like a sophisticated printer.

- Easier ways to search for needed information, even when you don't know what it's filed under or who has the file.

- Faster interdepartmental reporting.

- Greater availability of specialized software.

- Electronic mail functions to eliminate telephone tag and help ensure timely delivery of important messages.

Committee members should be encouraged to share this information with their constituents and to solicit reactions to it.

An Equipment Census

While they're surveying co-workers about possible LAN applications, your committee members can also be making a record of what kinds of electronic equipment are installed in your facility, how these devices are currently used and by whom. This is information you probably won't need until later, when you are talking with network vendors about your needs, but now is the most efficient time to gather it. Forms like those shown on the following pages will allow your committee members to record all the pertinent information.

A Technoliteracy Census

There's one more important information-gathering task to be performed before you tackle the process of deciding to LAN or not to LAN: You need to assess the levels of technophobia that exist in different areas of your company.

Your committee members will survey their constituents once more, asking non-threatening questions that will reveal how comfortable different kinds of workers are with technology. In other words, the committee members will not ask, "Are you frightened by all these computers?" but "How do you like this machine? Does it make your job easier? Was it hard to get used to it? If you could swap it for something else, what would you get?"

The answers to these questions tell you how comfortable your employees are with the technological tools they're using now—without making them feel like Bob Cratchits if they're *not* comfortable yet with those tools.

The need for this information should be obvious. You don't want to waste money installing a company-wide LAN if the workers who

Facsimile

Location: _____

Current Equipment

Manufacturer	Model #	# of Units	Lease	Own

Operators: # Day _____   # Evening _____   # Night _____

Hours in Operation: # Day _____   # Evening _____   # Night _____

Telephone Lines

Dedicated: # Day usage _____   # Night usage _____

Dial access: # Day usage _____   # Night usage _____

Messages

	Internal	Domestic	International
% Incoming			
% Outgoing			

Facsimile Locations

	Domestic	International
# Sent		
# Received		

Word Processing

Location: _____

Current Equipment

Manufacturer	Model #	# of Units	Lease	Own

Utilization

Internal _____

External Domestic Link-up _____

Ext. International Link-up _____

Operators: # Day _____   # Evening _____   # Night _____

Terminals: # Day _____   # Evening _____   # Night _____

Messages

	Internal	Domestic	International
% Incoming			
% Outgoing			

Usage Percentages

Text _____   Filing _____

Statistics _____   Correspondence _____

Charts/Graphs _____   Other _____

Future Plans for Expansion: _____

_____

_____

_____

Electronic Mail System

Location: _____

Current Equipment

Manufacturer	Model #	# of Units	Lease	Own

Software Used   _____   _____

                        _____   _____

                        _____   _____

Operators: # Day _____   # Evening _____   # Night _____

Hours in Operation: # Day _____   # Evening _____   # Night _____

No. Messages/Day	Internal	Domestic	International
% Incoming			
% Outgoing			

Projected future plans/growth: _____

_____

_____

_____

Other Office Systems

Location: _____

Current Equipment

Manufacturer	Model #	# of Units	Lease	Own

Utilization

Internal _____

External Domestic Link-up _____

Ext. International Link-up _____

Operators: # Day _____ # Evening _____ # Night _____

Terminals: # Day _____ # Evening _____ # Night _____

Future Plans for Expansion: _____

_____

_____

_____

Computers
Location: _____

Current Equipment

Manufacturer	Model #	# of Units	Lease	Own

Utilization

Speeds: _____

Codes: _____

Protocols: _____

Conversion: _____

_____

Operators: # Day _____   # Evening _____   # Night _____

Shifts: # Day _____   # Evening _____   # Night _____

Electronic Messaging	Internal	Domestic	International
Average cost			

Future Plans for Expansion: _____

_____

_____

_____

are carrying floppies around now are going to keep doing that, rather than learn to use the network to save time.

Another summarizing meeting with the committee now will enable you to winnow the list of *possible* applications to those:

- That offer the highest possible gains in efficiency.

- That offer new opportunities to the company.

- That offer the highest likelihood of being fully utilized.

Now you know **what you want your local area network to do**. If you like, at this point you can pare down the size of the committee, but keep the representatives-and-constituents structure; it will not only smooth the selection process, it will be a real boon if you do wind up installing a local area network. All your employees should understand that their comments and suggestions are welcome. All your committee members should understand that it is part of their committee-related job to relay comments and suggestions made to them

CHOOSING LAN PARAMETERS

The next question your committee will tackle is: **Which kind of LAN?** This is not as difficult as you (or your committee members) might think.

You'll start by outlining for the committee the seven characteristics that all local area networks have in common. All LANs are:

Local—up to 2-mile range
Private—user-owned and -controlled
Structured—pathway that connects independent devices
High-speed—1, 4, 10 Mbps
Commercially available—can be purchased
Transmitters of packets of data — for most efficient use of the network
Connectable—to link one device to others

It's important that you be the one to present this information to the committee. It establishes for all concerned the level of technological sophistication that is going to permeate the committee's work. It's a way of saying, "This is our starting point. To the extent that we learn to speak tech, we're going to learn it together."

If committee members have questions about generic networks at this point, it's okay to call on a technophile to answer them, but you must police the answers. If you don't understand, you can be sure the person who asked doesn't understand, but that person may be too embarrassed to say so. It's up to you to say, "I don't quite get it," or "Is this a question we should tackle later on, when we all know more about the subject?

DIFFERENTIATING LANS

Beyond those seven generic characteristics, there are four key factors that distinguish one local area network from another. Installing a LAN means making a choice in each of these four categories:

- Media

- Topology

- Access method

- Modulation

As you'll see in the next four chapters, the choice in each category is virtually dictated by:

- What you want the LAN to do.

- What you might want the LAN to do in two or three years.

- How user-friendly you need the LAN to be.

- What the physical limitations of your site are.

Chapters 4–7 explore each of these four key factors and relate them to the needs of different kinds of user organizations. You'll be surprised at how easy it is to choose the media, topology, access method and modulation that are right for your company.

CASE STUDY — Part III

To gather the sort of nitty-gritty details he wants about every aspect of his corporation, Sly Fachs has asked each of his department and division managers to serve on Electromedia's LAN committee and to bring two additional representatives—one middle manager and one clerical or technical employee. Thus, Electromedia's committee will span the breadth of all the company's operations and be able to gather information and ideas from every employee level.

For the initial, information-gathering phase, the committee will include 45 representatives, three each from:

The accounting department.
Each of three book-publishing divisions.
Each of six magazine divisions.
The circulation/fulfillment department.
The software company.
The direct-mail marketing division.
The typesetting and printing center.
The company-wide support services division (which includes human resources, building services, procurement and market research).

Although this is an almost-unwieldy number for committee meetings, the real work of this committee is not to assemble for meetings, but to gather information in discrete work environments. Once the information has been assembled, the committee will be pared down to 15 representatives, one from each department and division.

End of case study (Part III)

SECTION TWO:
WHICH KIND OF LAN?

Chapter 4

Media

If you were going to build a private road for vehicles to travel in and around your company's site, you'd think about how much traffic that road would need to accommodate and what kind of traffic — 18-wheelers, commercial vans, passenger cars, bicycles. You wouldn't spend the money to build a superhighway for a small amount of lightweight traffic. On the other hand, you wouldn't expect big trucks that visit your plant regularly to be able to get in and out on a bicycle path. Under certain circumstances, you might decide that a combination of roads is the most cost-efficient solution. This is what choosing LAN media is all about.

Each network medium available now has its own pluses and minuses, but a characteristic that's a plus from your company's point of view might be a minus to another prospective user. Therefore, you need to weigh each medium's characteristics against the identified needs of your company to get your money's worth out of a LAN.

There are two categories of LAN media: terrestrial and free-space. Terrestrial media include twisted-pair wiring, coaxial cable and fiberoptic cable. Free-space media, for use in highly specialized situations, include digital microwave radio and infrared light beam transmission.

TERRESTRIAL MEDIA

Twisted-Pair Wiring

Twisted-pair wiring is the simplest, least expensive medium available for a local area network. That's an advantage if you don't need (and therefore don't want to pay for) high capacity. If you anticipate that your network will be used by only a small number of employees, primarily for slow-speed (remember that "slow" is anything less than apparently instantaneous) applications like electronic mail, twisted-pair may be the right choice for your LAN or for part of your LAN.

On the other hand, if you install a twisted-pair LAN and try to send too much data along that network, you're going to have delays and traffic jams. Sooner or later, one of your 18-wheelers is going to bog down and miss a scheduled delivery. For a high-volume network, twisted-pair proves to be a false economy.

Twisted-pair wiring is a familiar sight in most offices. Pairs of plastic-sheathed twisted copper wire like these are commonly used to connect digital equipment and telephone extensions to PBX (switchboard) systems.

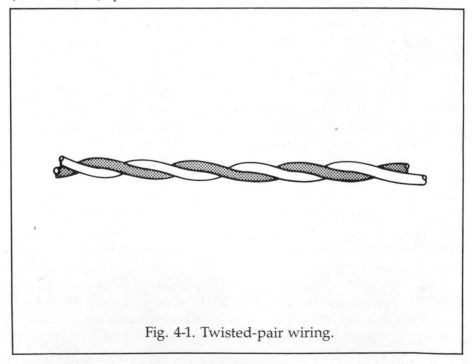

Fig. 4-1. Twisted-pair wiring.

In a local area network, twisted-pair can yield speeds of up to 1 megabit per second (Mbps). For comparison, the twisted-pair wiring installed with your telephone system is probably operating at slower speeds, from 9600 bps to about 64 Kbps. But remember that the typical speed range for LANs is 1 to 10 Mbps, which means twisted-pair is a relatively slow LAN medium.

A twisted pair LAN can theoretically support up to 1,000 devices spread over a distance of up to 15 miles, but most twisted-pair networks are much smaller. Since twisted-pair is slow and small, as LANs go, it's naturally the lowest-cost medium to install, ranging from about 25 cents to $1 per foot. In most situations, this will make a twisted-pair LAN your least-expensive option, but that may not be so in all cases. Depending on what kinds of devices you want to hook up to the network, and depending on how far apart those devices are, a twisted-pair LAN may require the installation of expensive special interfaces (on-ramps) and/or special boxes called repeaters that keep the signal from fading out when it has to travel a long way between one device and the next or to push the LAN to its highest achievable speed.

Another consideration about twisted-pair cable is that it has "low noise immunity"—it is subject to interference from other electrical or telephone wiring nearby, for example. In some cases, you can overcome the interference problem, if you need to, by running the twisted-pair cable, or certain lengths of it, through a shielded conduit. Naturally, that increases the cost of installing the LAN.

Conduit is also used in some twisted-pair installations to protect against accidental physical damage to the wire. You can see, if you look at some of the wires around your office, that twisted-pair is not as substantial or sturdy as some other kinds of cable. This is important if the wire in a high-traffic floor area can't be installed under the floor or in overhead conduit, or if part of the wire will be outdoors. Again, conduit pushes up the cost of installing twisted-pair.

One other thing you should know about twisted-pair wiring: It is the easiest of all LAN media to tap. That's a plus or a minus, depending on your company's needs. The word "tap" means exactly what you think it means, the same thing it means in all those detective movies on late-night TV: unauthorized listening. But here it also means authorized listening. If you want to hook a new computer up to an existing LAN, you will *tap* the LAN to create an on-ramp for the new device. If you want to hold down the cost of adding new devices to your LAN later, the ease and economy of tapping twisted-pair is a plus. If you intend to use your LAN to transmit sensitive data, the ease and economy of tapping twisted-pair is a minus. One way to improve the security of a twisted-pair LAN is to "shield" the cable—naturally, at higher cost. If you need both low expansion cost and high security, you may want to use a different medium for the portions of your LAN that are carrying sensitive data.

It is possible to mix media on a local area network. Most often, this is done by using a higher-capacity, higher-speed, more expensive medium as a sort of main street (a "riser" or "backbone") and less expensive media, including twisted-pair, for side streets.

Twisted-Pair Wiring Characteristics

- Inexpensive.

- Easy to install.

- Easy to tap.

- Low noise immunity.

- Speed of up to 4 Mbps.

- Can accommodate up to 1,000 devices.

- Adequate for network span up to ½ mile.

Coaxial Cable

By far the most popular medium for today's local area networks is **coaxial cable**—the same cord that brings cable TV into your home. Actually, there are many kinds of coaxial cable ("co-ax" to insiders). Some are designed for use with *baseband* systems, others for use in *broadband* networks—the subjects we'll look at in Chapter 7. Otherwise, the chief differences among various kinds of coaxial cable are in the materials used for the components of the cable.

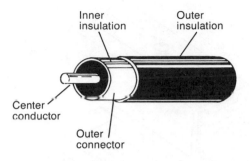

Fig. 4-2. Coaxial cable.

A center *conductor wire* — the axis that gives this cable its name — carries the transmission signal. It is *insulated* by nonconducting materials, such as PVC or Teflon, which are surrounded by another *conducting* material — usually solid copper or a woven mesh of fine copper wire — all encased in another layer of *insulation*, sometimes called a sleeve. This outer sleeve may be extruded aluminum.

Because coaxial cable offers higher immunity than twisted-pair wire to noise, it affords higher bandwidth (capacity) and higher speed. It costs three to four times as much to install this medium, before you add in the cost of required interfaces and other network components.

One advantage of coaxial cable is that it is widely used in non-LAN applications. This means it is widely available, which helps bring down the cost, and it relies on what is already a mature technology.

Maximum speeds achievable on coaxial cable range to 10 Mbps or more, depending on how much *bandwidth* is available — how much of the road surface can actually be used by vehicles. Bandwidth is affected by the kind of co-ax installed. Your choice — how much speed do you need? — depends entirely on how much traffic you're going to send and how fast it needs to get where it's going. If twisted-pair is emphatically too slow for your needs, you have a wide range of options in coaxial cable.

Coaxial cable is not as easy to tap as twisted-pair wire. Once again, this is a blessing or a curse, depending on your needs. Usually tapping a coaxial cable requires piercing and coring the cable and then clamping on a tap.

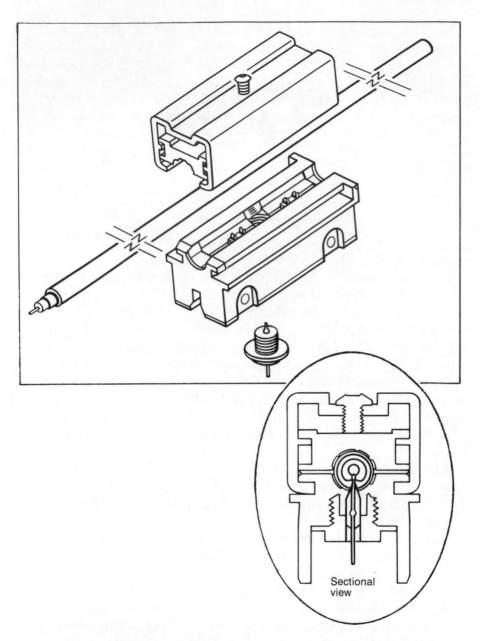

Fig. 4-3. Coaxial tap.

Another good news/bad news aspect of coaxial cable is that it offers much more potential for expansion than twisted-pair. Good news if you expect the kind of corporate growth that will increase your LAN usage over the next three to five years. Bad news—you're paying for a lot of underutilized capacity—if you don't expect (or if you don't achieve) that growth. This is why you must assess not only your current local communications needs but future needs as well. This is where the non-communication lists you and your committee made in Chapters 2 and 3 can help you. Which of the applications identified there could be scheduled for future implementation? Do those applications justify the additional cost of coaxial cable, even if twisted-pair would do just fine for now?

Coaxial Cable Characteristics

- Widely available.

- Good noise immunity, high usable bandwidth.

- Speed up to 10 Mbps.

- More difficult to tap than twisted-pair wiring.

- More expensive than twisted-pair wiring.

Fiberoptic Cable

Or do those applications point to a need for an even higher-capacity LAN? The highest-speed, highest-capacity medium available now for local area networks is **fiberoptic cable**. You've seen commercials about this state-of-the-art medium and the ways it's being used in long-distance telephone networks. It's the medium of the not-too-distant future for local area networks, too.

Fiberoptic cable is a lightweight, rugged, high-speed, high-security medium that transmits light pulses over glass or plastic fibers. Each highly refractive fiber is surrounded by *cladding* with slightly lower refractive properties to isolate one fiber from its neighbors. One or more clad fibers make up the *core* of the cable. A plastic *jacket* surrounds the core to protect the cable from environmental damage.

Fiberoptic Cable

There are three basic types of fiberoptic cable: multi-mode step-index, multimode graded-index and single mode.

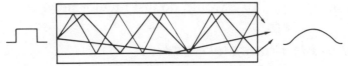

Fig. 4-4. Multimode step-index fiber.

Multimode step-index fiberoptic cable is the easiest to connect or splice and least expensive. It has a large core and high internal reflection properties. The glass cladding has a lower refractive index—this "step down" creates a boundary between the core and the cladding. Light pulses moving through the fiber hit the core's boundary at a glancing angle and are reflected back through the center of the core, toward the opposite boundary. While this propels the light signal forward, it also causes both pulse dispersion and spreading—it weakens the part of the signal that is moving forward. For this reason, multimode step-index fiber is normally used in short-distance, low-bandwidth networks.

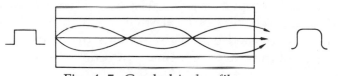

Fig. 4–5. Graded-index fiber.

Multimode graded-index fiberoptic cable offers comparatively higher bandwidth than step-index. The refractive index of the core gradually decreases from the cladding (where it's highest) to the core. This creates a more gradual boundary than in step-index, and causes less dispersion and spreading of the light signal. Multimode graded index fiber is most commonly used for long-distance telecom links.

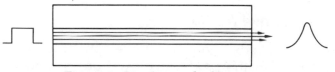

Fig. 4-6. Single-mode fiber.

Single-mode fiberoptic cable is somewhat more difficult to work with and more expensive than either of the multimode types. It has a smaller light-carrying core, which limits light pulse to a single path. This allows no variation, dispersion, or spreading of the light pulse.

Fiberoptic cable is still a more expensive medium than co-ax or twisted-pair, but costs are dropping rapidly, and some sources predict it will outpace coaxial cable as the LAN medium of choice before the end of this century. One significant difficulty with fiber today is tapping the cable. This problem will very likely disappear as fiberoptic technology matures, but it is one reason that the most common use of fiberoptic cable in LANs is as a riser—a main street that ties together clusters of local streets, which are themselves coaxial cable or twisted-pair. This obviates the need to tap the fiberoptic part of the LAN every time a new device is added to the network.

FREE-SPACE MEDIA

Fiberoptic cable, co-ax and twisted-pair are all **terrestrial media**. These are the pathways installed for most LANs, and they are likely to remain so. But other media, called **free-space media**, are also used in special situations. This is particularly true where the installation of a physical pathway—say, between two buildings where the appropriate right of way cannot be obtained—is impracticable.

Digital Microwave Radio

Digital microwave radio is one free-space option. It relies on radio technology to send the LAN signal from one station to another along a line of sight. This means there can't be any trees or other buildings in the way. Microwave transmissions are also subject to weather and other kinds of environmental interference, but they are the solution to some unique installation problems. Digital microwave radio should only be used to connect what otherwise would be inaccessible locations.

A microwave LAN, unlike the generic LANs described in Chapter 1, does require an FCC license.

Digital Microwave Characteristics

- Free-space medium.

- Transmits along paths in line-of-sight of each end.

- Subject to environmental interference.

- Rights of way not needed.

- Requires FCC license.

Infrared Light Beam Transmission

Infrared light beam transmission is the second-most popular free-space medium for a LAN. This sort of transmission is done with mirrors. Really. A laser or a light-emitting diode (LED) produces light pulses, which are received and or relayed by mirrors.

Like microwave, this medium relies on line-of-sight transmission. It is somewhat less vulnerable to weather than microwave, but still requires relay around trees, buildings, etc.

Infrared Light Beam Characteristics

- Free-space medium.

- Transmits along sight lines.

- Rights of way not needed.

- Less sensitive to weather interference than microwave radio.

Both microwave and infrared media have been used in experimental LAN installations, but neither is yet a mainstream medium. They seem to have the most potential for campus-like installations where running cable from building to building is either downright impossible or far too costly.

ZEROING IN ON OPTIONS

It's entirely possible that other LAN media will come along in the
next decade or so, but for now those are your options. You won't set-
tle on specifics until Chapter 8, but it's possible to narrow the choice
a bit now.

- What's the geographical scope of your LAN likely to be? If it's
  more than 10 miles, you won't want an all-twisted-pair LAN.

- How much traffic will your LAN carry? Is that enough to justify
  the cost of a fiberoptic riser?

- Will your LAN take in more than one building? If so, can some
  kind of physical cable be run between them? If not, you may
  want to consider a free-space medium for at least part of your
  LAN.

Take these questions up with your user committee. Don't worry
if the technophiles in the group are besotted with the Space Age
technology of fiberoptics and microwave transmission. If those
aren't the media that meet your company's needs, this is the discussion that will lay those considerations to rest.

Now that you've eliminated at least some potential building
materials for your road, it's time to look at how you want to lay that
road out, a question that may clarify your media options still further. The techspeak name for this LAN characteristic is "topology,"
and it's the subject of Chapter 5.

Chapter 5

Topology

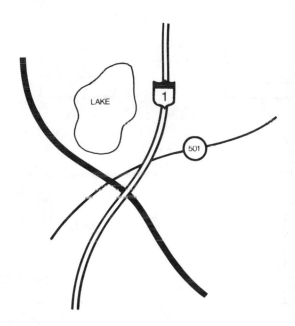

One of the most baffling questions about local area networks is: How did LAN creators come up with a name as fearsome as "topology" for what they might just as easily have called "floor plan" or "road map?" Look up "topology" in *your* dictionary, and see if the definition is any less intimidating than this one, from the *Random House College Dictionary*:

> **to-pol-o-gy**, *n. Math.* the study of the properties of geometric forms that remain invariant under certain transformations, as bending, stretching, etc. Also called **analysis situs**.

Sort of makes you feel local area networking is a decision best left in the hands of high-paid experts, doesn't it? In fact, of the four distinguishing characteristics of LANs, topology is the easiest to understand. What's more, since **any topology can work in any situation**, it's virtually impossible to make a wrong choice. This is a piece of cake in Sacher torte clothing.

BASIC TOPOLOGIES

Topology is the **spatial pattern formed by the physical links** of a
network.

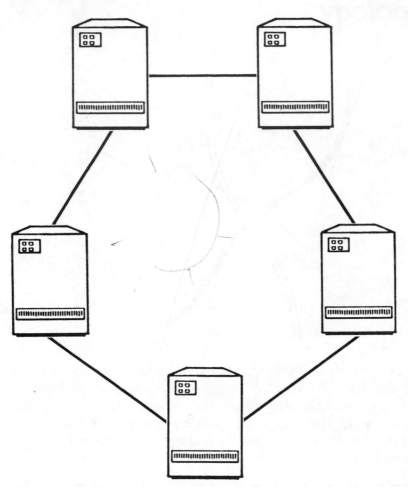

Fig. 5-1. Ring network.

The links may from a **ring**, so that data passes from one station
to the next, to the next and so on, returning eventually to the sending station. This is comparable to the beltway-style express roads
that surround some major cities. Since traffic around a ring network
is generally one-way, a ring doesn't necessarily yield the shortest
physical distance from sender to receiver, but because LANs operate at such high speeds, that's not a serious drawback (Fig. 5-1).

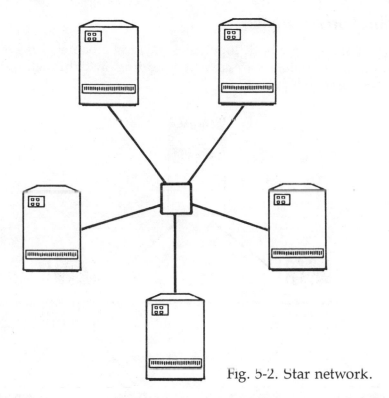

Fig. 5-2. Star network.

The links may form a **star**, so that all data passes from the sending station through a central point and on to the receiving station or stations (Fig. 5-2). This hub-and-spokes array is not unlike the complex freeway interchange in downtown Los Angeles. It's not the shortest route between Santa Ana and San Bernardino, but it is a usable option and under some conditions it may be the most efficient.

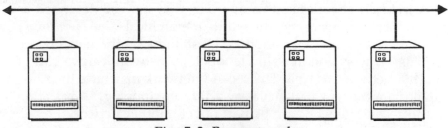

Fig. 5-3. Bus network.

The links may follow a back-and-forth route, like a shuttle **bus**. A transmission moves from one station to the next, until it reaches the end of the line, where it turns around and will pass the same stations again, in reverse order (Fig. 5-3).

Branched Networks

Those are the three basic topologies for local area networks. More complicated topologies are created by adding on to (**branching**) one of these simple topologies.

Fig. 5-4. Hybrid network.

A **hybrid** network like this one (Fig. 5-4) may be described as hierarchical. In this usage, the word "hierarchical" does not imply that one station outranks another; all stations on the network are "equals." Any station in a simple-topology network can be part of another network, as well. The second network may have the same topology as the first, or it may have a different topology. Any station in the second network can be part of yet another network.

If you try to diagram the topology of such a network, you'll wind up with something that looks like a **tree**—another name that is often listed as a basic LAN topology (Fig. 5-5). Usually the word tree is reserved to describe a branched LAN whose limbs are all bus-topology networks.

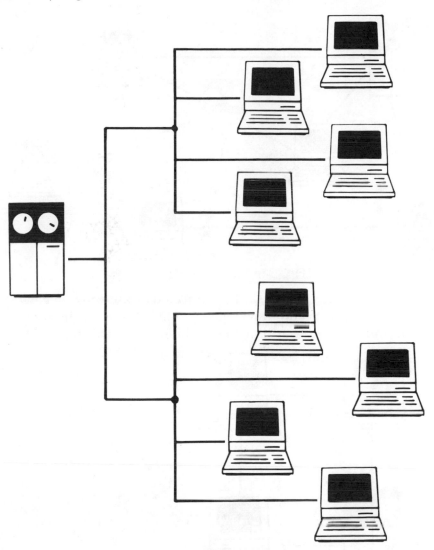

Fig. 5-5. Tree network.

These five words—**ring, star, bus,** and **hybrid** or **tree**—can be used to describe any LAN topology. Unfortunately, the same people who brought you the word "topology" have applied their skills to naming topologies, too. Add a marketing department's talents to the obfuscation, and you'll read and hear about topologies called *SABER* (Self-Amplifying Bus Extending Ring), *LINC*, *mesh*, *grid* and no doubt more to come. You can play this game, too. Here's something one network designer calls a mesh topology (Fig. 5-6):

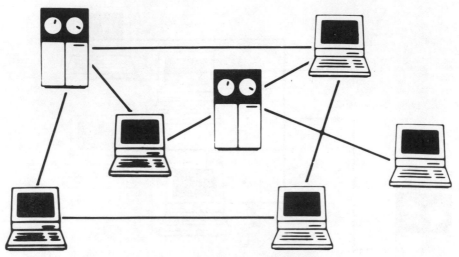

Fig. 5-6. Mesh topology.

Notice which stations are physically connected to one another. Now rearrange the stations and draw your own lines to make the same physical connections. Your drawing could look like this (Fig. 5-7):

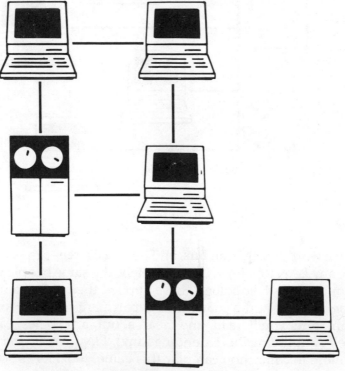

Fig. 5-7. Topology "X."

What does that look like to you? A "double loop with extended bus?" A "figure-eight?" Call it a "Graham cracker on a plate" if you like; you've just invented a new topology.

Virtual Pathways

Notice that any station on any network can communicate with any other station — one of the defining characteristics of all LANs. No matter how circuitous the **physical pathway** from one station to another might be, a direct, **virtual pathway** exists from every station on the network to every other station on the network. Notice that you would still have the same virtual pathways if you lined the seven stations up some other way and connected them in a ring, star, bus or hierarchical topology, instead of copying the physical pathways in the original "mesh" configuration. This is the reason any topology *can* work in any situation, though certain topologies are used most frequently in combination with particular media or access methods.

CHOOSING AN APPROPRIATE TOPOLOGY

For all the technobabble about LAN topologies, the simple truth is: Any topology — ring, bus, star, or hybrid — can be made to work in any company. If that's so, you might wonder, why aren't all networks set up in a ring, say, or a bus topology? Why bother with even as many as three basic topologies, not to mention all the branched hybrids they can yield? There are differences between one topology and another that may make one choice more efficient than another for your particular needs. Before we even look at those, there are two other considerations that may make the choice for you.

Physical Limitations of Your Site

In some buildings, it may be impossible or prohibitively expensive to run cable or conduit along any pathway but one—possibly a duct installed during construction to accommodate future needs for additional wiring. If so, the scope and placement of that duct will have a strong influence on your choice of topology. Ususally, it will point the way to a bus network as the least costly topology for at least that segment of the LAN.

What Your Vendor Has to Offer

If your company is already heavily committed to one supplier of electronic office equipment, it makes a lot of sense to talk to that same vendor about networking that equipment. This doesn't mean you can't or shouldn't talk to any other vendor, but you certainly won't overlook Wang's LAN offerings, for example, if nearly every machine you already own is a Wang product. After you've analyzed several appropriate alternatives, perhaps Wang's FastLAN will turn out to be big enough, fast enough, versatile enough and priced right for your needs. FastLAN is a bus-topology network.

DISTINGUISHING CHARACTERISTICS

If neither of those considerations simplifies the decision-making process for you, the characteristics that distinguish one topology from another become important. As with the choice of media, it's often a case of one user's meat is another's poison.

Ring Topology

The earliest commercial LANs were based on ring topology, a relatively simple arrangement and therefore a relatively economical approach. A ring network has no "controlling node": Every station on the network has equal responsibility for network control. That is, each station picks off any transmission addressed to it and passes all transmissions along to the next station in line. The next station repeats the process, but that is not the first station's problem. Traffic generally moves in one direction only, though it can be reversed if necessary.

One catch in this technique: If one station conks out, the circuit is broken and the whole network stops (Fig. 5-8). The network stops, too, if the circuit must be broken for the addition of a new station. There are solutions to this drawback, but they will add to the price of the network.

One solution is to add to the network the programming to recognize that a transmission it is trying to pass along isn't being received at one station, and to pass the transmission in the opposite direction instead (Fig. 5-9). This converts the ring to a back-and-forth U shape — you might picture it as a bus that happens to be curved — until the disabled station can be brought back to life. In

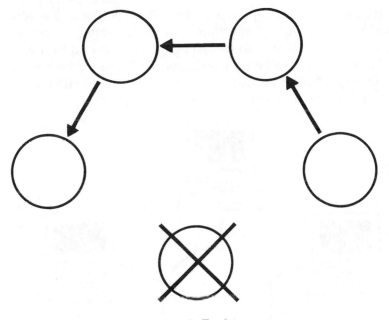

Fig. 5-8. Problem.

other ring networks, a physical pathway exists around every station, so traffic can detour around an out-of-service computer or terminal.

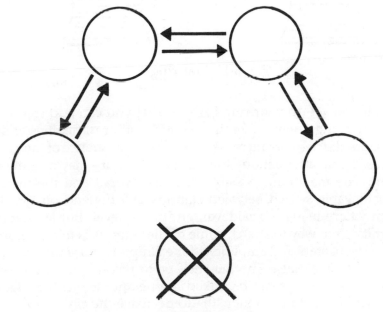

Fig. 5-9. One solution.

Another solution is "bypass circuitry." A LAN like this is some-times called a "star ring." Each transmitting/receiving node on the ring passes a transmission to its respective station and to the next node. If a station is disabled, the node will still pass transmissions around the ring. Naturally, this arrangement, with its extra wire and extra hardware, is more expensive than a simple ring (Fig. 5-10).

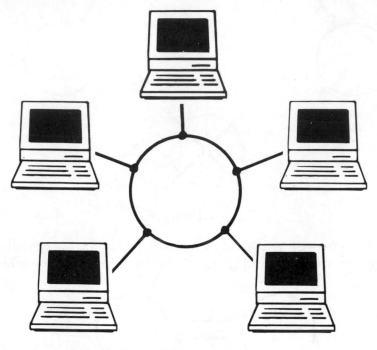

Fig. 5-10. Star ring.

Even if you set up your ring LAN so that you can add stations without bringing down the entire network, other difficulties arise as the size of the network increases. As the circumference of the ring increases, signal attenuation—the signal fades out—may become a problem. The maximum possible distance overall and the maxi-mum achievable stretch between stations will depend on which medium you're using, signal level, and noise level, but all media have limits. One way to stretch some of these limits is to add more hardware (repeaters and amplifiers, for example) whose function is to boost fading signals. Another may be to reconfigure the LAN —several smaller rings can be linked to one another with bridges (which we'll look at in Chapter 10)—to decrease the circumference of any one ring.

As the number of stations on a ring LAN increases, response time increases also: It takes longer for a transmission to be passed along by 200 stations than by twenty, so it will take longer for your request to reach the database, and longer for the database to send back the file you asked for. This is another judgment call: How fast do you need the network response time to be? In some situations, two seconds will be much too slow. In others, 4.5 seconds will be just fine. Only you and your network users can answer this question, after you've figured out what tasks you want to carry out via the LAN.

Ring topology would seem to be most appropriate for relatively small, stable-size networks or for small branches of a hybrid LAN. Its relatively low cost makes it attractive, too, for companies that want to tiptoe into networking—"Let's see if this works in one department before we take it company-wide." It may not be such a good choice for a network that will encompass hundreds of stations, extend over a great distance or need frequent expansion.

Star Topology

A star-topology network relies on point-to-point transmission (in either direction) between a central host and each station. The host station acts as a clearinghouse that reroutes station-to-station transmissions. Generally, the software that controls the network resides in the host, too. This means the host station must either be dedicated to doing nothing but running the network, or it must have enough spare processing and memory capacity to operate and control the network without sacrificing whatever other jobs it has. The host of a star-topology LAN may be part of a computerized PBX ("private branch exchange" or switchboard). LAN capability is now offered as an option on several business telephone equipment systems.

What all that points to is the use of a star topology in a network where most of the traffic runs from the host—say, a mainframe computer or database machine—to one or more stations and back again, rather than a network with lots of station-to-station traffic. Such a LAN functions just like the old mainframe-plus-terminals arrangements you've had in your data processing department for years, but now there is a virtual pathway between terminals as well.

The capacity of the host is the largest single cost factor in installing a star network. Initial installation costs are generally higher for a star network than for other topologies. However, the traffic load along any single physical pathway will always be a fraction of the total network load, so low-speed, low-capacity (low-cost) media will often be adequate for the job. Where traffic problems are most likely to occur in a star network is in the host. In that case, network overload points to a need to install a still larger host—say good-bye to the money you saved installing less-expensive wire if you didn't plan for this contingency.

If your host is large enough to accommodate more stations, expanding a star network is relatively easy and inexpensive. The network need not be disabled to install a new station, and the incremental cost of expanding is comparatively low. The distance between the host and a station is limited only by the medium used. The medium will also dictate the maximum network speed achievable.

The maximum network speed achievable, you should be aware, is not necessarily the speed you will get on any network. This is particularly true on a star network, where all station-to-station traffic is momentarily stopped while the host reads the address and then reroutes the transmission. This is what network designers mean when they talk about **throughput**. The freeway analogy applies here, too: Just because the speed limit is 55 per hour, that doesn't necessarily mean you can travel 55 miles from your home in 60 minutes. Detours, bottlenecks, and heavy traffic can all slow you down on the freeway (Fig. 5-11).

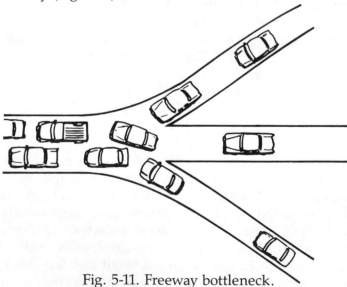

Fig. 5-11. Freeway bottleneck.

Since every station-to-station physical pathway in a star network is a detour, and since all traffic must pass through the bottleneck of the host, station-to-station throughput will necessarily be substantially slower than the maximum speed offered by the medium you've installed. In fact, data throughput will be lower than raw speed in *all* local area networks, for a variety of reasons. In a star network, the topology itself is one of the reasons.

Note, however, that because the route from station to host is free of competing traffic, detours, and bottlenecks, station-to-host throughput in a star network can be almost as high as the raw speed offered by the medium, assuming that the host can handle it and other similar connections simultaneously. If your company's needs for a LAN rest heavily on communication between a host and satellite stations, with minimal requirements for station-to-station communications, a star topology may be an appropriate, even optimum, choice. This is especially true if the host, with spare capacity, is already in place in your company and you're looking for a way to let more users access the information and processing power of that machine.

One other factor may make a star-topology LAN particularly attractive for your company: If you've already installed a telephone system that is designed to deliver LAN services (as an option), it will almost certainly be cheaper to enhance your phone system by adding that option than to install a separate local area network. Generally, activating the LAN capabilities of a telephone system that has such an option requires the addition of some new software and circuit boards to the PBX (the refrigerator-size boxes that fill a room in the basement); some new wiring will be required to connect individual computers and terminals to the network, but the main wiring arteries, between your PBX and various offices in your building, will already be in place. If your company's PBX is a relatively new one, it may have a LAN option. Ask your telecommunications manager. If your company is contemplating the purchase of a new telephone system, you may want to consider a system with this option.

Bus Topology

Bus topology always seems to come in second when you compare it to ring and star on any particular grounds: cost, reliability, flexibility, ease of expansion, maximum scope. . . . Perhaps that's why it's coming in first as the most popular overall choice, the Miss Congeniality of topologies.

Unlike a ring network, where all traffic travels one way around the loop, or a star, where traffic moves first one way (from station to host) and then the other (host to station), traffic on a bus network travels two ways simultaneously.

In most bus networks, each station **broadcasts** its transmission to *both* of its next-door neighbors. Each receiving station will pick off the transmissions addressed to it and rebroadcast all transmissions (so that a transmission intended for several stations is not short-stopped by the first one to receive it).

There obviously has to be a way to keep all this traffic from colliding. In fact, there are several ways to do this. One way is to lay parallel wires, separate traffic "lanes," along the bus route. Naturally, this multiplies the cost of the cable. Another is to build traffic control into one end of the network so that a host or network controller limits access to the network. Obviously, this reduces overall throughput by lengthening the time a transmission must wait to get on the bus.

A third option is to use **broadband** coaxial cable—a single cable that can be divided into separate channels. We'll talk in Chapter 7 about **modulation**, the technique that allows a single cable to be divided into channels this way. If you've already guessed that a multi-channel network is more expensive than the first two alternatives, you're getting the idea.

If a station in a bus network conks out, it won't disable the entire network, as it would in the simplest rings. The signal from station Number One will sail right on by disabled station Number Two and be received by station Number Three. Even if the transmission medium itself is physically damaged, some part of the network will continue to function until repairs can be made.

Throughput may decline as stations are added to a bus topology. Depending on the access method used (see Chapter 6), either the response time will worsen as stations are added (each station will have to wait longer for a bus to arrive) or heavy traffic may create problems (each station will have to wait longer for a bus with empty seats).

That very difficulty gave rise to the versatile tree topology—a bus that has been branched into several smaller buses. By isolating clusters of stations in this way, the tree configuaration can reduce traffic congestion in a large network, maximize throughput, and minimize systemwide downtime. Branching a ring network generally requires shutting down the whole network, a serious disadvantage once you've come to rely on your LAN. Star networks are easier to branch than rings, but have other disadvantages (including higher initial cost). Finishing second in the ease-of-branching race is one factor that has made the bus topology the most popular starting point for LAN users who expect to grow.

NETWORK COMPONENTS

All these topologies were named for the overall shape of the **network link**—the express-road part of the network. At least, ring, star, bus and tree are what those network links looked like to the people who named them. Naturally, there's more to this data transportation system than the express road itself. There are surface streets and access roads (on- and offramps). These are officially called **links** and **interfaces**. In some cases, they are already built in to data equipment — you won't see a separate box or length of cable that eases data into the flow of traffic on the network link. When they're not invisible, this is how they are arranged (Fig. 5-12):

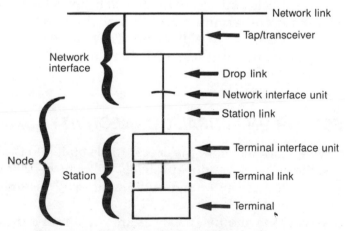

Fig. 5-12. Network links and interfaces.

Whether or not you ever see these gizmos, chances are you will hear about them, possibly from someone who wants you to believe this whole subject is much too complex for you. By now you should know better.

The segment called **network interface** is the connection between each node and the network link. Everything coming onto the network must be compatible with the network link, so if the signal from the node is not compatible, the **network interface** must translate or convert it. In a **distributed control** network like a ring, where each station has some responsibility for keeping the network running, this signal conversion is done by the **network interface unit** (NIU). Besides the NIU, the network interface includes the **tap** or **transceiver**, which diverts traffic onto and off the network link, and the **drop link**, the cable that carries data between the tap and the NIU. In a star network with a centralized network controller, there are no NIUs and no drop links.

Everything on the *other* side of the network interface makes up the **node**, which consists of the **station** and the **station link**, the cable that connects the station to the network interface. These two will almost always be visible to you. If there's no NIU (and no drop link), the wire you see running from the tap to your computer is the station link. Your computer or work station is the station. It may have a built-in **terminal interface unit** (TIU), which adapts the signal from your terminal to the network interface. If the TIU *isn't* built in, you'll see that box and a cable—the **terminal link**—running between it and your **terminal** (computer, word processor, printer).

These surface streets and access roads will be invisible to you if they are built in to your terminals or if they are built in to the network. They *can* be built in to either one. They can be built in to neither. There are names for these variations, too.

THE QUEST FOR UNIVERSAL CONNECTIVITY

A **proprietary** network will have the access roads built into the network. That is, the network vendor—probably the same vendor who supplied your computers and other equipment—will assume the responsibility for building a network all those devices can use to communicate with one another. Chances are you'll find it difficult or impossible to connect equipment from other hardware vendors to this network. Not so long ago, computer vendors thought this was a good way to 'lock in' their customers.

Not too surprisingly, some other computer vendors saw proprietary networks as a way of locking them *out* of the marketplace. This gave rise to the **standardized** network. By far the best-known standardized network is Ethernet, developed in a collaborative effort of more than a dozen vendors, headed by Xerox, starting in 1981. A standardized network is an "open" system—its specifications are not a closely held secret; the responsibility for building devices that can communicate on the network rests with the hardware manufacturers. Not every manufacturer chooses to participate in this kind of networking. Even if they did, there remains the question of *which* network standards will a vendor subscribe to?

The Holy Grail of networking—the **universal** network that will allow any data device to communicate with any other data device as easily as your electric typewriter draws current from the wall —remains an elusive but hotly pursued goal for vendors and users alike. We'll examine this subject more closely in Chapter 11. For the time being, at least, networking equipment from different vendors means creating a wide variety of ways to move data onto and off the expressway.

There's more to getting on that expressway than interfaces. Network access can be the equivalent of trying to make a left turn at a busy intersection during rush hour. To manage that problem, network designers have devised a variety of traffic-control methods called access methods, the subject of Chapter 6.

Chapter 6

Access Methods

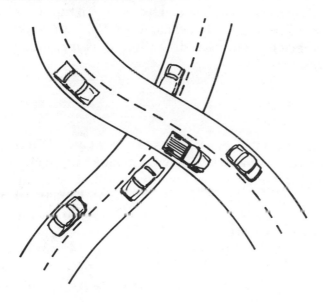

Although local area network access methods can be compared to the etiquette drivers ought to observe when entering an expressway, it's simpler to visualize them in terms of the way traffic is controlled at the intersection of two surface streets.

ALOHA

Where two lightly traveled roads intersect, there may be no stop sign or traffic signal at all, and regular drivers along both roads develop the habit of breezing on through the intersection. This is roughly comparable to the earliest network access method, known as ALOHA. As long as traffic on the network is a once-in-a-while occurrence, this can work.

Occasionally, however, there will be a **collision** that will stop both transmissions from reaching their destinations. In the case of highway traffic, the result may be death and destruction. In the case of network traffic, the result is an inconvenience that forces both transmitting stations to try again. On pure ALOHA, there was no way of knowing that a collision had occurred; users came to call it the "transmit and pray" system. As traffic on the network increases, naturally the potential for collisions grows. Before long the need to retransmit data stopped by collisions becomes burdensome. In fact, a network needn't be carrying very much traffic at all for this point to be reached, and that is the reason ALOHA is not a feasible access method for today's local area network. Even though ALOHA is not offered on commercially available LANs, it does provide a frame of reference for understanding the access methods that are used in today's networks.

CSMA

After a sufficient number of accidents occurs at an ALOHA cross-roads, local drivers learn to slow down and look out for cross traffic before entering the intersection. The townsfolk may petition local authorities to install "Yield" signs. The comparable network access method was originally called Carrier Sense Multiple Access. Nowadays, it's known as **CSMA**.

In a CSMA network, the station that wishes to transmit 'listens' first for other traffic on the network. If none is detected, transmission can begin. If traffic is detected, the station waits a short time, then listens again for a clear path.

CSMA/CD

While CSMA, like Yield signs at an intersection, reduces the chance that a collision will occur, it doesn't eliminate that chance. If two or more stations hear the same clear path and begin transmitting at the same time, there will be a collision. As the number of transmissions on a network increases, so does the chance that this will happen. To solve that problem, an improved version of CSMA evolved: **CSMA/CD**, or Carrier Sense Multiple Access/Collision Detection (Fig. 6-1).

With CSMA/CD, the transmitting station continues to listen to the channel while it is transmitting. If the station detects a collision (a higher than normal voltage swing), it stops transmitting and waits briefly before beginning the listen-then-send process all over again.

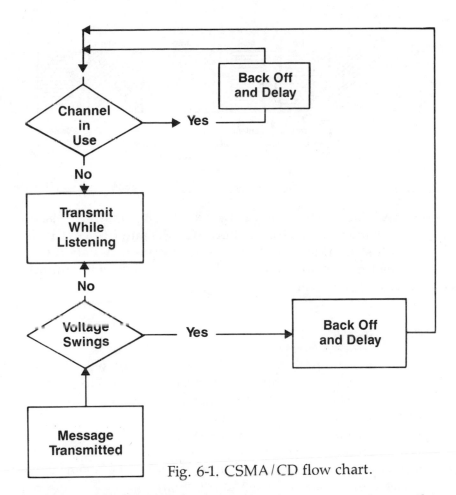

Fig. 6-1. CSMA/CD flow chart.

Collision detection becomes more and more necessary as the number of transmissions on a network increases. Without it, collisions will result in neither of the affected transmissions getting through—and the sending stations will not know their transmissions didn't get through unless some other method of verification is used. The verification process uses some of the network's transmission capacity, reducing useful throughput (Fig. 6-2)

The disadvantage of CSMA/CD is analogous to a conversation where you "can't get a word in edgewise." Lengthy transmissions, like Station Number 1's in Figure 6-2, keep the network channel tied up. Plain bad luck can keep one station off the network if there's a high volume of traffic. This is what's happening to Station Number 3 in Figure 6-2.

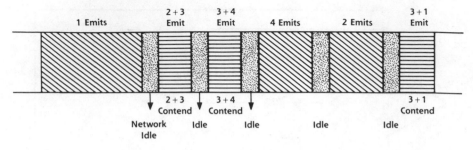

Fig. 6-2. CSMA/CD traffic pattern.

One way to get around the "can't get a word in edgewise" disadvantages of **contention** access methods (CSMA and CSMA/CD) is to give each station, in turn, a chance to transmit. A station that doesn't need to transmit can "pass" until the next time around. Three access methods are based on this "in turn" concept.

TOKEN-PASSING

Token-passing is roughly comparable to the right of way that applies at a four-way stop. The car that stopped first gets to proceed first. When that car has cleared the intersection, then it's the turn of the car that arrived at its stop second, and so on. Although the danger of collision is almost nil, every car has to stop at this intersection, so traffic is slowed down somewhat. Similarly, every station in a token-passing network must wait for the token – and wait longer if many stations wish to transmit.

In a token-passing network a "token" (bit signal) is passed from node to node. A station that wants to transmit "takes" the token in its turn and does not pass the token to the next station until after it has sent its own message. Token passing is associated with ring and bus topology networks (Fig. 6-3).

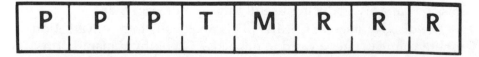

Fig. 6-3. Token. P = priority bit;
T = token bit; M = monitor bit;
R = reservation bit.

The "token" is actually a message itself, which moves from station to station. The token bit (T in Figure 6-3) may be 0 (free) or 1 (busy). The remaining bits serve different functions, depending on network vendor. The monitor bit is responsible for the "welfare of the network"—policing, for example, to be sure there is only one token circulating on the network at any one time (Fig. 6-4).

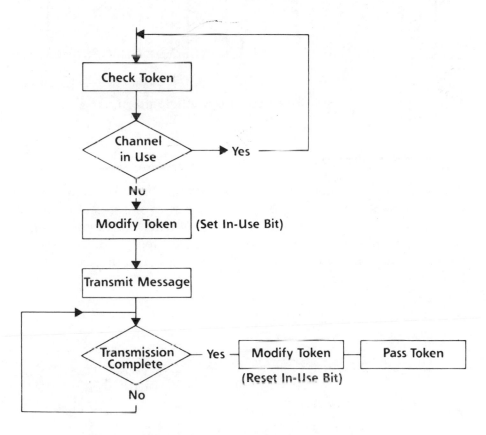

Fig. 6-4. Token-passing flow chart.

If the token is free, a station wishing to transmit seizes it, changes the designation of the token bit to busy, and then transmits. When it has finished transmitting, the station modifies the token again, returning it to free status, and passes it to the next station (Fig. 6-5).

Token passing permits each transmitting station to retain control of the network for as long as necessary to transmit an entire message. It ensures that each station will receive the same number of opportunities to transmit. That number decreases as the number of stations using the network increases, and the passing of the token itself uses up some of the transmission capacity of the network.

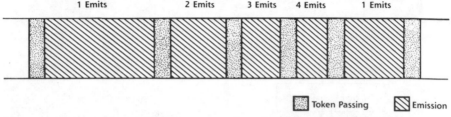

Fig. 6-5. Token-passing traffic pattern.

SLOTTED ACCESS

Slotted Access, another "in turn" access method, can be compared to a traffic signal. A green light allows traffic from one street to enter the intersection. An entire funeral cortège or truck convoy may not get through the intersection with the same green signal, and a single car may have to wait through several green-red cycles if it's behind a lot of others.

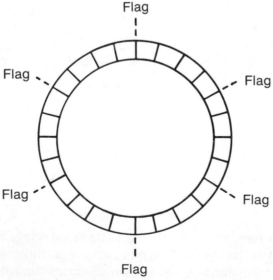

Fig. 6-6. Slotted ring.

In a slotted-access system, data-packet windows ("slots") circulate from node to node to be filled with packets from transmitting stations. Each slot is preceded by a "flag," which, like the token in token passing, is changed from free to busy status by the using station (Fig. 6-6).

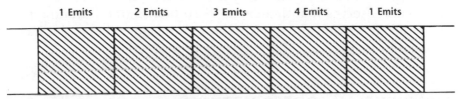

Fig. 6-7. Slotted access.

Note that all the slots in one network are the same size (Fig. 6-7). Messages too long for a slot must be broken into shorter segments or packets for transmission. Slotted access is most often used in ring topology, but may be used in other topologies. It is most useful in a network where nearly all transmissions will be of the same length. This allows the slot size to be established so as to leave a minimum of wasted space, without forcing users to break messages into pieces more than occasionally.

Both token-passing and slotted-access networks rely on the stations in the network to pass access from one to another. This is one reason these access methods are popular in **distributed control** networks like the ring, where each station is a more or less equal partner in keeping the whole network running.

In a **centrally controlled** network, where network responsibility resides in a single host or controlling node, it makes more sense to let that node function the way a traffic cop would at that intersection we've been visualizing. When there's no traffic in the north-south lanes, the cop allows east-west traffic to flow without interruption.

STATION POLLING

The networking equivalent of that traffic cop is an access method called station polling. Station polling resembles token passing in its effect: Each station gets the opportunity to transmit in turn and may transmit a message of any length before relinquishing that opportunity. This opportunity is offered to each station by the host, rather than passing from station to station.

Like a traffic cop, the station-polling host may offer permission to transmit **sequentially** (now it's Station A's turn, even if Station A is only going to transmit the message that it has nothing to transmit). Obviously, this wastes a certain amount of network time, but remember that the network operates at such high speeds that the delay may well be invisible to network users. Where even that delay would be intolerable, a system may use **roll-call** station polling: The traffic cop observes that there is no northbound traffic waiting to enter the intersection and quickly skips to the lane where a car has signaled by its presence that it wishes to proceed.

CHOOSING AN ACCESS METHOD

No one access method is inherently best for all applications. The choice depends on:

- Overall traffic volume.

- How often each node needs to transmit.

- How long each transmission is likely to be.

- Cost.

Since you will most likely want to use your LAN for more than one application, a compromise will often be required. You may have one application ideally suited to CSMA/CD (short, bursty transmissions) and another that points to token passing (many lengthy transmissions), for example. Thus, it's important to look at *all* the applications your LAN is likely to serve and to prioritize them before one access method is selected as the one that offers optimum efficiency for your operation.

Achieving optimum efficiency for your operation is the objective in selecting all three of the distinguishing characteristics of LANs we've examined so far. Identifying the right combination of medium, topology, and access method—the combination matched to your operation, your traffic volume, your need to expand—will point you to the local area network that really serves your needs. The fourth distinguishing characteristic—modulation—can actually *multiply* the capacity of your LAN. You'll see how in Chapter 7.

Chapter 7

Modulation

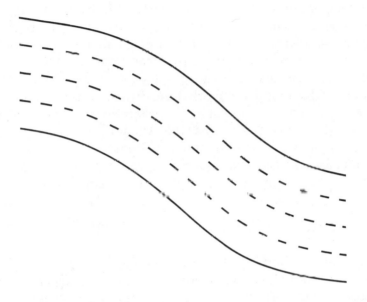

The buzzwords—you've heard them—are *baseband* and *broadband*. Quite simply, they refer to the number of *channels* a local area network has. Baseband is a one-channel system, a one-lane express road. Broadband is a multi-channel system, multi-lane highway —with one important difference from the roadway you're visualizing: No random lane changes allowed. In that respect, it is most like, and most often compared to, your cable television service. When you tune in Channel 7, you won't be interrupted periodically with sound and pictures from another station.

As fast and efficient as a baseband system can be, it may not be capable of carrying all the data volume your organization needs. One way to lick this problem would be to install additional, parallel baseband systems. This is every bit as inelegant and costly as it sounds, but it would yield the desired result: *greater aggregate throughput*.

On the other hand, if you know from the start that you want to transmit more data than a single baseband system can handle, why not build one multi-lane highway? Why not choose a broadband system and take advantage of the construction economies of building one wider roadway?

The baseband-broadband choice is essentially a matter of economy, but it's a bit trickier than road-building. Logic tells you the wider roadway ought to cost more to construct than the narrower one. Generally, the total cost of a broadband system will be higher than the total cost of a baseband system, but it's not because broadband coaxial cable is more expensive than baseband. In fact, broadband cable often costs less, per running foot, than baseband. The law of supply and demand is at work here. Broadband coaxial cable is the same cable used in your cable TV system so it's widely available for use in LAN installations as well. What raises the cost of a broadband system is the additional hardware required to *modulate* the data signal for *analog transmission* from point to point. Broadband systems also require more complex design engineering.

ANALOG LIVES

Yes, "analog transmission" is what you read there, and if you found that surprising—after all you've heard and read about *digital* being the technology of the future and analog becoming something of a dinosaur—you've just stumbled onto an ironic truth. Old technologies never die, they just get applied to different problems. Good old analog technology is what makes broadband systems—the superhighways of the digital LAN world—possible.

As we saw in Chapter 1, digital transmission is a highly efficient way of moving data from one piece of digital equipment to another. Depending on the medium used, hundreds of millions of data bits per second can move along a cable, with no slowdowns at either end for conversion to or from analog form. But they must move *single-file*, and there is a calculable limit to the total number of bits that can be transmitted along any medium during any particular length of time. When that limit is reached, the question becomes, "How could we make these data bits travel side-by-side and thus multiply the number that can pass over the medium in the same time interval?"

For a way to divide a single cable into non-interfering channels, LAN developers reached back to analog technology. Every medium used in local area networks has a comparatively maximum *usable bandwidth*—a range of frequencies at which data may be transmitted (Fig. 7-1).

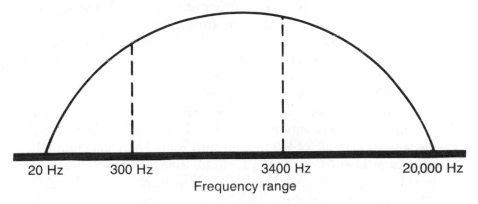

| 20 Hz | 300 Hz | 3400 Hz | 20,000 Hz |

Frequency range

Fig. 7-1. Available bandwidth.

Simpler media—twisted pair cable and *baseband coaxial cable* are, in effect, narrower roadways than broadband cable; only single-file digital transmission is practical over these narrow-bandwidth media (Fig. 7-2).

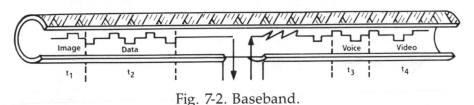

Fig. 7-2. Baseband.

BROADBAND SYSTEMS

Broadband cable, as the name implies, offers a wider roadway, but it cannot be physically partitioned off the way a road would be. Instead, the total frequency range is divided into channels or bands, just as it is in the coaxial cable that delivers your cable TV service (Fig. 7-3).

Fig. 7-3. Broadband.

Each channel in a broadband system is separated from the adjacent channel by a *guard band*—a frequency range over which no data will be transmitted. You can see this effect on your cable-connected television set or video cassette recorder at home: With the selector set on, say, Channel 7, turn the fine-tuning knob in one direction until you are pulling in a station you know is not Channel 7. Notice that the image on your screen does not slide from Channel 7 to the next station; you'll get a pattern of "snow" between the two. What you are seeing is a guard band in your cable TV system.

Depending on where you live, your cable television system may deliver more than 100 stations. Put another way, the bandwidth of that system is capable of being split into more than 100 channels, plus the guard bands needed to separate them. The previous illustration shows a broadband cable split into three channels (three different frequencies), but it is possible to create more or fewer splits.

Setting Up Channels

The simplest broadband LANs will be split into two channels—not necessarily of equal size—one for inbound data, one for outbound. Each of these gross channels may itself be split into two or many more channels (Fig. 7-4).

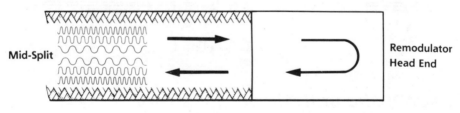

Fig. 7-4.

The total bandwidth in a single commercially available cable is 400 megaherz (MHz)—400 million cycles per second. In a *mid-split* system, frequencies from 5 to 108 MHz are reserved for inbound signals, frequencies from 162 to 395 MHz are reserved for outbound signals. Frequencies from 108 to 162 MHz are reserved as a guard band. The highest and lowest frequencies are generally not used because they are susceptible to noise and interference.

Broadband frequencies need not be split down the middle. A system may be *sub-split*, with, for example, frequencies from 5 to 30

MHz reserved for inbound traffic, and frequencies from 54 to 395 MHz reserved for outbound. Sub-split systems are most commonly used to accommodate heavy video traffic. Full-motion video is the only LAN application that requires an exclusive channel and thus, a broadband system. Freeze-frame video (series of still pictures) can be interspersed with data or other traffic in the single-file transmissions of a baseband system. Full-motion video can be transmitted on a baseband system if it is the only traffic on the system. However, if full-motion video is the only traffic to be carried on a system, closed-circuit TV will be a better choice than a LAN.

Each channel in a broadband system may be designated to carry a specific kind of traffic—video only, for example—or the network may be programmed to select an appropriate channel for each transmission. In either case, channel selection is done by the equipment, without user intervention. Neither the sender nor the receiver needs to "tune in" to a particular channel for a message to find its way from station A to station B. (A family of users "resides" on a specific broadband channel. A member of that family can join another family by changing its modem frequency.) This means that, even though a broadband system is usually far more complex than a baseband installation, it is no more difficult for humans to use.

The Role of Modems

Broadband systems are generally more expensive than baseband networks—not, as noted, because the cable is more expensive, but because of the additional hardware needed to convert digital signals to analog and back again. The device that does this job is called an *RF modem* (radio frequency modulator/demodulator). The modem receives a digital signal from the sending station, modulates it to the appropriate analog signal for the channel on which it will travel and sends it along the network to another modem, at the receiving end. The second modem remodulates the signal, making it digital again, to pass it along to the intended receiving station. Thus, a modem must be placed between every station on the network and the network itself.

It is possible for two or more stations to share one modem (Fig. 7-5), which lowers the cost somewhat compared with one-modem-per-station. Stations sharing a modem all need to be relatively light users of the network, or you get a traffic jam at the shared modem.

Heavy-use stations ("data hogs") need their own ("dedicated")
modems.

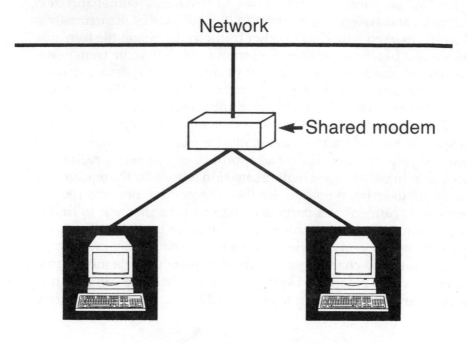

Fig. 7-5. Shared modem.

Once a signal is transmitted by the sending modem, it must stay
on its own channel—in its own lane of traffic—until it reaches its destination modem. To change lanes—to move from an inbound lane
to an outbound one, for example, the signal needs to be remodulated: Its frequency must be increased or decreased to match the second channel. The box that does this job is called a *remodulator* or,
sometimes, a "head end" (from CATV technology).

Other Hardware

Another piece of hardware that is unique to broadband systems is
the *directional coupler* or "signal splitter." This may be invisible, built
into the modems and head ends, but it's there nevertheless, doing
the job of selecting transmission channels and assigning frequencies.

Still other pieces of broadband hardware have baseband counterparts that go by different names. For example, all systems need devices that prevent fading signals ("signal attenuation"). In a broadband system, the job is done by *amplifiers*—which boost the signal level. Amplifiers also boost the level of network noise and interference and can actually garble the signal if they are not properly spaced throughout the network. In baseband systems, signal attenuation is reduced by *repeaters* – which actually regenerate (transmit a duplicate of the original) the network signal. Repeaters can filter out some excess noise and interference as they pass the signal along.

HOW DO YOU CHOOSE?

Now that you know the meaning of those two buzzwords—broadband and baseband—how do you choose the modulation scheme you need? There are really only two critical variables here.

1. Will you need to use your LAN for full-motion video as well as other applications? If the answer is yes, you must choose broadband. Now reread the question this way: Will you need to *use your LAN* for full-motion video applications? Remember, you don't have to put those applications on your LAN. It may be more cost-effective to install a baseband LAN and a separate, closed-circuit video system.

2. What volume of data do you expect to transmit on your LAN? If a single high-speed channel will be adequate to meet your local data-transmission needs over the next five years, you don't need broadband.

The noncritical variable in this equation is: How important is it for your company to be using state-of-the-art technology? This is a question that will come up as you and your user committee begin to wrestle with the task of matching the needs you identified in Section One to the local area network possibilities we've explored in Section Two. The matching process begins in the next chapter.

SECTION THREE:
WHOSE LAN IS IT, ANYWAY?

Chapter 8

Choosing a LAN

As you read Section Two, you may have been thinking about which
LAN variables would be right for your organization. You may have
decided, by the time you got to Modulation, that there are too many
options and that the process of choosing is too complex. In truth,
you've already done the hardest part of the job — needs analysis
— in Section One. Matching your needs to the appropriate LAN
characteristics becomes a fairly simple series of discrete steps, if
you've analyzed your company needs carefully to begin with.

That's a bigger 'if' for some organizations than for others, but you
and your committee should pause here to consider whether the
needs analysis you've already performed is adequate for you to proceed. If it isn't, call a time-out now and fill in the gaps.

In particular, make sure you have considered the functions and
work tasks of every employee who will use the network. That
means every employee who is equipped with a terminal, personal
computer, word processor, or other electronic station that could be
hooked up to the network now or later. Yes, every one of them, not
just their managers, and not just the data processing department.
If you really believe the DP manager knows precisely what the public relations director's secretary does, ask them both and compare
their lists. Ask the public relations director, too, while you're at it.
The results will demonstrate clearly why you must ask the individual employees themselves. Make sure you have done this before
you proceed with trying to match user needs to LAN capabilities,
or you'll find yourself inventing a cure for which there is no disease.

IDENTIFYING LAN APPLICATIONS

From the individual needs worksheets you completed in Section
One (or during that time-out), you can construct an aggregate picture of communications in your organization. Not all of these communications need (or ought) to be transferred to your network, of
course. It's time to separate the likely LAN applications from the
rest.

Use the LAN:	Do it the old way:
Access another computer's data base	Telephone and intercom communications
Interoffice memos	Face-to-face conversation
Share applications software	Circulating newspaper and magazine articles
Share in-progress data files	
New data-processing links	

You'll find a few communications tasks that could go in either category. Company-wide announcements, for example, could be disseminated via the LAN if every employee will be connected to it. If you're going to have to distribute hard-copy versions of broadcast memos to some employees, because they're not on the network, putting this function on the LAN creates more work, not less. The decision on how to categorize such a task depends on how 'fully electronic' your company is.

Another either/or example: In-house transmission of external communications. 'Carbon copies' of outgoing letters that were created on a networked computer or word processor can be transmitted electronically to in-house destinations, of course. But it makes no sense to re-keystroke an incoming letter just to share it with one or more colleagues. That's what photocopiers and office messengers are for. On the other hand, if you already own a scanner, communicating copier, or facsimile machine that can be tied to your network, electronic distribution of such materials is an option worth considering.

A healthy dose of common sense is called for in categorizing communication tasks. A committee member who's enchanted with state-of-the-art technology may suggest that 'while you were out' telephone messages be transmitted by LAN. Any employee who takes those messages regularly can tell you the familiar pink pads are a more efficient medium for any message of less than interoffice-memo length. All the more reason to have some nontechnical people—even a certifiable technophobe or two—on your committee.

HOW MUCH AND HOW MANY?

Once you've identified the tasks your network can improve, you want to figure out how many employees are involved in those tasks and try to predict the volume of traffic your LAN will need to carry. You may find it helpful to take a highlight marker to your needs-analysis worksheets now.

Obviously, everyone involved in the tasks you will use the network for needs to be equipped with a terminal. If this need is not already met, make plans now to acquire terminals or personal computers or other devices for the employees who don't already have them. You may want to postpone the actual purchase of extra equipment, however, until you've narrowed down your network-purchase options a bit. Nevertheless, count those to-be-acquired terminals in your census of machines to be connected to the network.

Arithmetic

To attach a meaningful number to anticipated traffic volume, you'll need to arrive at some lowest common denominator—a way to describe all the network traffic so that it can be totaled. For example:

- How many characters does a 300-page printout equal?

- How many characters in a dozen interoffice memos?

It's not your problem to translate the traffic into bits or bytes, but you need to quantify *all* the traffic in a uniform way: as characters, words, printed pages—square inches will do if that's what works best for you. Whatever your lowest common denominator is, how many of those units will you need to transmit in the course of a week? A day? A peak-traffic interval? No calculus training is required for this job, either.

Now it's time to translate those numbers into bits—or, more likely, megabits—per second. One of the technical wizards on your committee can do that, or you can ask a consultant or network vendor for help. For a trouble-free network, the number you want to calculate is: How many megabits per second (Mbps) must your network transmit during the busiest times?

Future Needs

The number may be surprisingly low, but we're not finished with the arithmetic yet. If you build your network to the speed specification you just identified, you won't have room to expand—to add other users to the network as your company grows or to add other applications to the network as your needs grow. This is where the "possible future needs" you hypothesized about in Section One come into play. If you factor that growth into your plans now, you'll choose a higher-speed network that you need today, and you'll be able to expand more cheaply when you need to.

Remember: The network speed you select in this way is primarily a reflection of the volume of traffic your network must carry, not (as the name would seem to imply) of how fast you need to communicate between stations. Now that you have this number, you're ready to begin choosing the LAN variables you want.

FOUR CRITICAL VARIABLES

The four variables that will define your LAN, and the choices in each category are:

Modulation	Media	Access Method	Topology
Baseband	Twisted-pair	CSMA/CD	Ring
Broadband	Coaxial cable	Token-passing	Star
	Fiberoptic	Slotted access	Bus
	Microwave	Station polling	Hybrid
	Laser		

THE BASEBAND/BROADBAND CHOICE

You must know the speed you want, the number of stations you expect to network, and the size of the geographic area your network will cover before you can make an informed choice between baseband and broadband. Once you know those numbers, the choice is sinfully easy. You need a broadband system if, including room to expand:

- you need a network speed greater than 10 Mbps

- the number of stations to be linked on the network is more than 500

- the geographic scope of the network will exceed 2 miles

- you need to transmit full-motion video images on the network

You may want a broadband system if your network will be placed in an environment with lots of electrical interference ("noise"). The coaxial cable used in broadband installations generally has much higher "noise immunity" than other media, but there are baseband media (such as Ethernet coaxial cable) that can conquer most noise problems. It is possible to "shield" (for example, run through conduit) simpler cabling for installations where noise is a problem.

If none of your organization's needs demand broadband, the choice is baseband.

CHOOSING MEDIA

If you chose a broadband network, you already chose your medium: broadband coaxial cable.

If broadband isn't required, it would appear that you have three media options. If fact, however, fiberoptic cable is rarely installed as the *only* medium in a local area network. It is frequently used in main trunks or backbones that carry network traffic from one department to another, and if your network will take in such a span, you may want fiberoptic cable for part of it.

The same properties that make fiberoptic cable a less-than-ideal choice for network-wide use are what make it a good choice for main trunks or "risers": It's relatively expensive to install (but risers require comparatively short lengths of cable) and it's more difficult to tap than twisted pair or coaxial cable (but risers need to be tapped only once at each department).

Even though it is not a broadband medium, fiberoptic can carry more than one network channel through a single cable; this is achieved by running more than one fiber through the cable, rather than by splitting the transmission signal into frequencies. Thus, it can yield a higher *aggregate* transmission speed than other media. With two fibers, each yielding a maximum speed of 10 Mbps, a fiberoptic riser can move data at a rate of 20 Mbps from one subnetwork to another.

With or without fiberoptic cable, for most baseband installations you will need to choose between twisted-pair cable and baseband coaxial cable. Two of the same numbers that helped you decide between baseband and broadband will answer this question for you, as well. You need baseband coaxial cable if, including room to expand:

- you need a network speed higher than 4 Mbps

- the geographic scope of your network will be one mile or more.

You may want to choose baseband co-ax if you think, hope or expect your company will grow faster and larger than you've projected. Technically, both twisted-pair and baseband co-ax can accommodate up to around 1,000 stations, but the more stations you have on your network, the more traffic you'll have. To accommodate all that traffic volume without intolerable slowdowns, you'll probably need the higher speed baseband coaxial cable offers.

Baseband co-ax also offers better noise immunity than twisted-pair. If potential electrical interference is a consideration at your site—but not so high as to have forced you to choose a broadband system—you may want to choose baseband co-ax for that reason alone.

The table on page 103 summarizes the key features of various LAN media.

CHOOSING THE ACCESS METHOD

If you've already decided you need a broadband system, your access-method choice is made for you: A broadband system requires a centralized head end, and your access method will therefore most likely be station polling. Otherwise, you're not taking optimal advantage of the centralized network control functions built into the head end. Bear in mind, however, that your broadband network can be a superhighway that links smaller subnetworks, which could be broadband systems. Choosing broadband doesn't get you entirely off the hook on this decision-making chore.

Choosing the access method for your baseband network is somewhat more subjective than picking baseband and media. The question here is, What kind of traffic will be traveling on your network? You'll need to go back to your need-analysis worksheets and think about these questions:

- How many stations are on the LAN?

- What's the length (in characters, words or some other common denominator measurement) of a single network transmission?

- How often does each station need to transmit?

Comparing the Major Media

	Twisted pair wire	Baseband coaxial cable	Broadband coaxial cable	Fiberoptic cable
Topologies supported	Ring, star, bus, tree	Bus, tree, ring	Bus, tree	Ring, star, tree
Maximum number of nodes per network	Generally, up to 1024	Generally, up to 1024	Generally, up to about 25,000	Generally, up to 1024
Maximum geographical	3 kilometers	10 kilometers	50 kilometers	10 kilometers and up
Type of signal	Single-channel, uni-directional, analog or digital, depending on type of modulation used; half- or full-duplex	Single channel, bidirectional, digital, half-duplex	Multi-channel, unidirectional, RF analog, half-duplex (full-duplex can be achieved by using two channels or two cables	One single-channel, unidirectional, half duplex, signal-encoded lightbeam per fiber; multiple fibers per cable; full-duplex can be achieved by using two fibers
Maximum bandwidth	Generally, up to 4 Mbps	Generally, up to 10 Mbps	Up to 400 MHz (aggregate total)	Up to 50 Mbps in 10 kilometer range; up to 1 Gbps in experimental tests
Major advantages	Low cost May be existing plant; no rewiring needed; very easy to install	Low maintenance cost Simple to install and tap	Supports voice, data, and video applications simultaneously Better immunity to noise and interference than baseband More flexible topology (branching tree) Rugged, durable equipment, needs no conduit Tolerates 100% bandwidth loading Uses off-the-shelf industry-standard CATV components	Supports voice, data, and video applications simultaneously Immunity to noise, crosstalk, and electrical interference Very high bandwidth Highly secure Low signal loss Low weight/diameter; can be installed in small spaces Durable under adverse temperature, chemical, and radiation conditions
Major disadvantages	High error rates at higher speeds Limited bandwidth Low immunity to noise and crosstalk Difficult to maintain/troubleshoot Lacks physical ruggedness; requires conduits, trenches, or ducts	Lower noise immunity than broadband (can be improved by the use of filters, special cable, and other means) Bandwidth can carry only about 40% load to remain stable Limited distance and topology Conduit required for hostile environments	High maintenance costs More difficult to install and tap than baseband RF modems required at each user station; modems are expensive and limit the user device's transmission rate; complex initial engineering	Very high cost, but declining Requires skilled installation and maintenance personnel Experimental technology; limited commercial availability Currently limited to point-to-point connections

Reprinted with permission from Datapro Research, Delran, New Jersey.

In most companies planning for a LAN, these "traffic loading" questions won't be easy to answer. Some users may never need to transmit anything longer than a one-page memo. Others, particularly the data-processing department, may need to transmit hundreds of pages' worth of information at a time. Some stations may need frequent access to the network, while others may need access less than once an hour. Nevertheless, you must answer these traffic loading questions or your LAN will turn into a traffic nightmare.

What you want to come up with is a sort of picture, like the crossroads we described in Chapter 6.

If the network traffic will be "short and bursty" — automobiles from all four directions arriving at the crossroads from time to time, but rarely at the same time and none of them taking very long to get through the intersection (Fig. 8-1) — a contention access method (CSMA or CSMA/CD) will work just fine. It may work, too, if

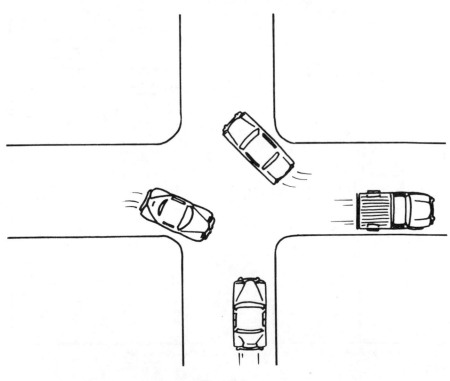

Fig. 8-1.

lengthy transmissions (a funeral cortège moving through the intersection) are a rare occurrence. Short and bursty traffic is most commonly found on personal computer networks; as soon as you put a minicomputer or a mainframe on the network, you're looking at lots of funeral cortèges needing to get through that intersection and holding up all the other traffic while they do.

Short and bursty, but heavier traffic—more automobiles arriving at the crossroads at the same time and thus a greater need to take turns with the right of way (Fig. 8-2)—calls for a less random method of control. Token-passing, a method that allows each station regular opportunities, in turn, to transmit, will work for this kind of traffic in the same way that a four-way stop works at an intersection. The token allows each station to transmit a message of any length during its turn, so an occasional behemoth can be accommodated without depriving other stations of their turns for an intolerably long time.

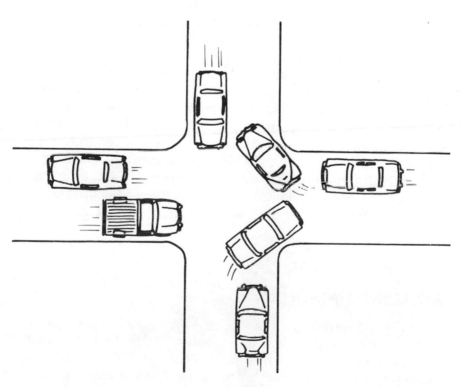

Fig. 8-2.

The responsibility for passing the token to the next station rests with each station on the network; there is no centralized control in a token-passing network. This makes it a popular choice for personal-computer networks.

Where nearly all transmissions are approximately the same length, slotted access — a traffic light that gives each direction an equal interval of 'green light'—can be used. Exceptionally long transmissions, which won't fit in the slot of time available, can be broken into smaller segments and transmitted piecemeal in packets. In some systems, the size of the slot can be changed ("dynamically reconfigured") from time to time to accommodate longer transmissions. Like token-passing, slotted access is a 'distributed control' method; no one computer on the network controls the movement of time slots from station to station. In a personal-computer network where each station needs approximately the same number of opportunities to transmit messages of approximately the same length, this method can be more efficient than token-passing.

Token-passing and slotted access systems don't work quite so well when many stations are transmitting very lengthy messages. Picture three funeral cortèges beating you, alone in your car, to a four-way stop (Fig. 8-3). Such systems lose efficiency, too, in offering frequent transmission opportunities to stations that need to transmit only infrequently. Picture yourself stopped at a red light when there's no traffic for miles in any other direction. A traffic cop could look at either of these situations and make a logical decision about the most efficient way to move traffic through the intersection. Generally, a network that includes one or more large computers will have such a traffic cop—a centralized network controller. Among its functions will be station polling: It will ask each station in turn (some stations may be allowed more turns than others), "Do you need to transmit now?" and allow traffic to flow efficiently.

CHOOSING A TOPOLOGY

The good news about choosing a topology for your network—or for subnetworks in a hierarchical arrangement — is that any topology will work. *You can't make a wrong decision here.* Still, you do have to make *some* decision and there are some factors that will point you most logically in one direction or another.

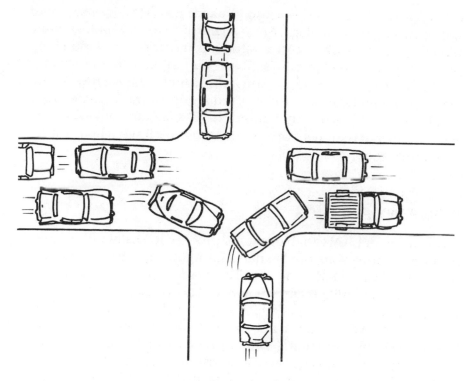

Fig. 8-3.

Topology is one case where having chosen broadband doesn't nail down your choice completely, though broadband networks are most often configured as bus or branched tree (hierarchies of buses) networks.

In baseband networks, the ring topology is most often used with token passing or slotted access, bus topology with token passing or CSMA/CD, and star topology with station polling, but other combinations are possible, and may even be desirable depending on your company's needs.

It is not uncommon, for example, for a company to want to use wiring that is already in place—sometimes as part of the telephone system—as the medium for all or part of a LAN. The topology defined by that in-place wiring may be the best topology for you to choose, if only because of the money you save by not rewiring. Rest assured that you will not make major network efficiency sacrifices by basing your topology choice on such a consideration.

Each topology matches up best with one access method or another—star with station polling, bus with token passing or CSMA/CD, etc.—in ways that are fairly easy to visualize if you think again about the vehicular traffic analogy. Since all traffic in a star topology must pass through a central point, a traffic cop at that point (station polling) is an eminently logical way to prevent intolerable traffic jams. Token passing and CSMA/CD on a ring topology have their counterparts in the way cars move into and around a traffic circle. Most LAN vendors have already figured these matchups out—and factored them into their network offerings, as you'll see when you solicit information from suppliers.

COMPANY POLITICS

Choosing your LAN variables should proceed logically, almost easily, if you've done your needs-analysis homework to begin with, but it doesn't always work that way. The hardest part may be getting your user committee members to agree on what those individual needs-analysis worksheets add up to. Which functions need the LAN most? ("Mine.") If the network is to be phased in, which department should get it first? ("Mine.") Whose needs ought to get priority in the compromises that will almost certainly need to be made, particularly on the access-method decision? ("Mine.") Once the network is in place, which department 'owns' and maintains and controls it? ("Mine.")

The power plays that can develop at this stage—even by committee members who were certifiable technophobes when you started this chapter—may amaze and confound you. Nevertheless, you're the referee, and it's up to you to distinguish between legitimate concerns about access, user training, network maintenance and efficiency, on the one hand, and plain old power grabs on the other.

A Company-Wide Resource

Information is a resource that belongs to the company, not to the department that creates, catalogues, updates or stores it, and the network that moves that information around must be treated the same way—as a company-wide resource—if its potential is not to be wasted. You'll want to keep that in mind, and you'll want your vendor and all your employees to keep it in mind, too, as you move on to choosing a vendor and implementing your network in the next chapter.

CASE STUDY—Part IV

Electromedia Inc. won't be able to use a local area network to estab-
lish the kind of link it needs to control its distant software subsidiary. That's a concern that will have to wait, for now. However, its
typesetting and printing subsidiary, only two miles away, is within
the range of a LAN if appropriate rights-of-way can be obtained to
extend the medium from Nyack headquarters to West Nyack. Substantial economies and efficiencies can be achieved with such a link,
not only for the magazine- and book-publishing operations, but
even for the high-volume word processing function that is handled
by the typesetting center.

In addition, the department and division managers have identified numerous ways to improve their own efficiency and profit planning if they (but only they) can tap the company's primary
minicomputer in the accounting department.

There is somewhat less urgency to make electronic mail available
to mid- and lower-level employees in the company, but that is a
LAN function the managers will certainly want to see phased in
later, when every company employee has access to a computer or
terminal of some kind.

The aggregate speed required for the first phase of this network
(the Nyack-West Nyack link and the mini-to-managers links) is cal-
culated to be about 3 Mbps. If a link to the Cambridge software lab
can be added later, if all employees are tied into the system for (at
least) electronic mail, and if the number of employees continues to
increase at a rate of 5 percent a year, the committee anticipates that
the system speed required will remain under 10 Mbps for at least
five years.

Therefore, a broadband system does not seem to be required
here. Since part of the network will need to link a remote (Nyack)
site, and the medium may be exposed to weather as well as electrical
interference, coaxial cable—possibly shielded co-ax for the length of
any outdoor runs—is a better choice than twisted-pair for Electromedia. Fiberoptic cable will be used for the longest stretch, between
Nyack and West Nyack. The accounting department's data processing guru unsuccessfully pled a case for a fiberoptic riser between the
two headquarters' floors. Ultimately, though, the other managers
—and, more important, CEO Sly Fachs—dismissed this as another
case of data processing's love affair with state-of-the-art technology.

For the initial phase, at least, CSMA/CD will be an utterly adequate access method, teamed with a branched tree topology. The

managers will be linked to the main mini via a bus network, which will branch at the minicomputer housed in West Nyack. When mid- and lower-level employees are added to the system, those stretches of cable may be branched at each profit center. This will keep intra-profit-center electronic mail from taking up space on inter-departmental stretches of the network.

End of case study (Part IV)

Chapter 9

Implementing a LAN

Network vendors can supply useful information to you and your LAN committee as you attempt to identify the most pressing LAN applications in your company. You may have talked with several by the time you narrow down the list of LAN characteristics you want. If that's the case, then you're already on the prospect list of those vendor representatives. Indeed, you may have acquired more product literature than you know what to do with—not to mention the regular phone calls you've been getting. If not, it's time for you to make some phone calls, and invite some vendor reps in to talk with you and your committee.

A meeting with your committee is a good way to start with each representative whose company offers a LAN that meets the basic criteria you settled on in Chapter 8. Your committee members—or an 'executive committee' of selected full-committee members—may raise issues with vendor representatives that you hadn't thought of, or that aren't significant to you personally but that will be of critical importance to one or more work groups in your organization. In addition, these early meetings will set a democratic tone for the implementation process that is quite different from the let-them-eat-LANs atmosphere that may prevail when a vendor has to 'sell' its LAN to only one key executive or manager, without worrying about how (or if) the network will be used by other work groups. You'll get a better network, and a smoother implementation, if your vendor understands from the start what you envision as the ultimate scope of your LAN. It's *important* that you keep your committee involved after needs identification, through vendor selection, installation, and expansion.

PRODUCT DIFFERENTIATION CONSIDERATIONS

In theory, at least, the network you've defined in terms of modulation, medium, topology, access method and number of stations to be served ought to cost the same, no matter which vendor sells it to you. Naturally, that won't be the case at all. Now you're ready to look at the product-differentiation messages of LAN vendors and consider the various ways in which different vendors add value to their network offerings. One or more of your committee members will want to know about the following factors:

Service and Support

What's the failure rate (mean time between failure) of networks the vendor already has installed? To what extent will the vendor 'be there' for you when the network malfunctions? When you expand your offices or move to a new location? When you need to add stations? When you need to add a different brand of stations than you started out with? How much does after-sale service and support cost?

User-Friendliness

Nearly all local area networks operate "transparent to the user"—the user doesn't need to be a technician to send or receive data from another network station. Nevertheless, the user does have to give the system some commands and instructions. How complex are they? How resistant will various employees in your organization be to learning what they need to know to use the network effectively?

Staff Training

Will the vendor train your staff to use the network? All your staff? Key operators? How much staff time will need to be devoted to training? What other customers can you call to find out how much time their training took? What training materials and other resources will the vendor provide? Will the vendor train, or provide training materials for, your new hires? At what price?

Ease of Expansion

How complicated, and how expensive, will it be to add stations to the network? To add equipment from different vendors? To link the LAN to a larger network that reaches your outlying offices or suppliers' or customers' offices?

Price Discounts

What can the vendor do to hold down your installation costs? Your expansion costs? If there is wiring in your building that can be used for the LAN, will the vendor use it? Will doing so wind up costing you more money than installing a network 'from scratch?' How much will it save you? Can the vendor pass along to you discounts on any network components? Would you save money by buying directly from a manufacturer? What's the trade-off (in service, support, training, etc.) if you opt for such a plan? Can you save money by "beta-testing" a new network product – that is, by being the guinea pig who helps the vendor identify problems with the system before it's been installed at dozens of sites?

"What I Want to Know . . ."

You should be familiar with the most important category of questions your committee members will ask vendor representatives. These are the questions that tell the vendor what you want to *do* with your network. That is the information the vendor needs to plan the most useful, most effective LAN for your company. "Will I be able to use someone else's spreadsheet software on my computer?" "Will I be able to work from home?" "Will someone else be able to read my confidential files?"

It's important for you and your committee to have such questions – even half-baked ones – answered. It's important for your vendor to know exactly what kind of performance you expect from your network. These are the exchanges that give everyone the information they need.

SELECTING THE GRAVY

If you've found six vendors who can all give you the baseband co-ax, tree topology, CSMA/CD network you want, the factors that differentiate one LAN from another are, in a sense, gravy. But one combination of flavors is likely to be more appealing to you than another. That is what you're looking for now: The best gravy for *your* company's meat and potatoes.

In the end, of course, the choice of vendor will depend on the price—not just who offered the lowest price, but who will deliver the most of what you need for the lowest price? That's a decision you may not want to make democratically, but you can make a better decision, and perhaps avoid costly mistakes, if you use your committee to help you elicit all the pertinent information about what each vendor is offering and about how important those factors are to your company.

Service and support are likely to be a major consideration for any LAN purchaser.

User-friendliness, on the other hand, may not be as important to you as it would be to other prospective customers. If your business is in a technology-related industry, and all your staffers are familiar and comfortable with computers and other electronic equipment, this needn't be high on your list of concerns.

Staff training may be important to you even if user-friendliness isn't—particularly if you are installing a broadband network that will require at least one in-house network manager to supervise it. On the other hand, if you already have a superlative in-house technical training staff, vendor-supplied training is likely to be of less value to you.

Ease of expansion probably will be an important concern, though your company may anticipate a different kind of expansion than another organization: You may be interested in the ease of adding in-house stations to your local area network, while another company might be considering future needs to communicate with outlying offices.

Price discounts are important to everybody. Unfortunately, they are too often important because of what was lost when the company 'bought cheap.' If you agree to be the beta-test site for a local area network, you can save lots of money, but be aware that the system just might fail at a critical moment. You may want to hang up a sign that says, "Every silver lining has a cloud." That's not to say you should ignore opportunities to save money, only that you must be realistic: If you need a failure-proof network for critical applications, if you need a high-capacity network, if you need a high-security network, *don't* opt for discounts that put your network's ability to meet those needs in jeopardy, but *do* skip the optional wood-grain finish on the network control console.

HOW DO WE PLUG IT IN?

Having involved your committee in vendor selection will pay off at implementation time with a team spirit that minimizes confusion, grumbling, roadblocks, and downtime. You and your committee should have impressed your priorities and concerns upon the vendor so the vendor's installation team arrives prepared to meet your expectations—from reassuring the technophobes to testing the system to be certain it does all that it's supposed to.

In preparing for implementation, your committee will have carried appropriate news about the network back to the people who are going to use it. End users needn't be unpleasantly surprised with interruptions necessary for installation or training. They will be prepared to ask the questions they need answered and to answer questions they're asked by the installers. They will know they should give your committee members feedback about the system, so that any problems can be diagnosed and solved quickly. You can expect one or two of them to discover applications you hadn't thought of, despite all your planning.

This kind of feedback is most useful—most easily assimilated and acted upon—if you are phasing in installation of your LAN. The feedback you get from the first segment installed can help you head off problems in subsequent phases. It can help you revise training procedures as necessary for succeeding groups of your staff. It can pave the way for more and better teamwork as the later phases of the implementation occur; once the Phase One employees begin talking to others about the benefits the network has brought to their desktops, employees scheduled for later installation will be even more eager to be networked themselves.

Phasing in the LAN

Phased-in installation isn't possible in all situations, and even where it is, you can't always start quite as small as you'd like. Nevertheless, it's an important subject to consider with your committee. If we're going to phase the network in, where do we start? What's the smallest useful increment to begin with—one that's large enough to give us a true picture of system performance, but small enough to keep any problems that arise well contained? What's the next logical group to expand to? And the next?

Your vendor can help you with these decisions—particularly with identifying a work group that is large enough to offer a valid test of the system but small enough to make problems and benefits easy to identify for appropriate follow-up. Your vendor can also advise you about how a phased-in installation will affect your costs. You can expect the vendor to charge you more if you require 20 installation visits for a 200-station network than if you need only 10, and that may affect your phase-in plans.

TAKING THE NETWORK PULSE

You'll want at least two post-installation committee meetings—more if you're phasing your system in. The first, soon after installation begins, should concentrate on identifying any problems and finding solutions for them. Obviously, system malfunctions need to be reported to the vendor, and the vendor needs to solve them.

Other system difficulties may signal inadequate user training; some or all of your users may need additional training. Perhaps the initial training process was flawed and needs to be revised before

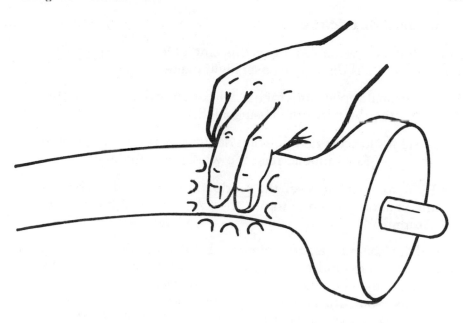

other employees are trained. This can be a touchy situation, particularly if frustrated system users are insensitively accused of "operator error" before the possibility of system malfunction is thoroughly investigated and eliminated. If users are having difficulty with the system and nobody on your own staff can identify the "errors" they are making, then someone from the vendor's staff should be available to walk through network procedures with the users as many times as necessary until he or she can say, "Aha! *That's* what you forgot to do!" or "Well, you're doing what you're supposed to, so there must be something wrong with the system." It is not too much to expect this kind of support from your vendor.

After your first post-installation committee meeting, you'll want your committee members to keep tabs on user progress in using the system: Difficulties? Complaints? New applications discovered? System shortcuts? Breaches of security? The most efficient way for you to find out about any of them is to make committee members responsible for bringing them back to the network user committee for discussion, resolution, celebration, broadcast—whatever's appropriate. This kind of feedback will streamline a phased-in installation, and will help your company get the most of its network after implementation is complete.

Additional Applications

One of the most valuable returns this kind of follow-up provides is the serendipitous discoveries users will make:

- Additional applications for the network that nobody thought of until after the network was installed.

- System shortcuts no one will know about until somebody on your staff finds time to read the whole user manual.

- System capabilities even the vendor didn't know about because the vendor never had to use the LAN in a real-business environment.

And don't be surprised if, before long, someone on your committee says, "If we could expand the network to reach outside this building or this company, we could. . . ." The fact is, smart, loyal employees like yours do appreciate the efficiency a network brings to their daily work lives, and it does make them think about other efficiencies that could be brought to bear on your company's business. So the question you want to think about, even before you install the LAN, is: "After the LAN, what?" Making your local area network one corner of a larger mosaic is the subject of Chapter 10.

CASE STUDY—Part V

Even though Electromedia Inc. is a publisher of technical magazines and books, less than one-third of the staff is technically oriented. Many who will need to use the network are technophobes, so user-friendliness is a big concern, according to committee members. This is a disappointment to Sly Fachs, who had hoped to train the staff by passing out user manuals and telling everyone to read them, rather than use company time to train network users.

Actually, Sly had considered his company a great candidate for a beta-test site, but his mangers have convinced him that a network failure will jeopardize on-time publication of books and magazines. They have also convinced him that a beta-site network will fail at some point.

The committee did come up with two cost-saving compromises, however. To save on providing user-friendly training to Electromedia's staff, one of Electromedia's own talented writers will produce an easy-to-use set of network instructions, no more than six pages long. Sly convinced one of the non-beta vendors that the lack of such simple instructions was a serious drawback to his system, and got the vendor to offer a discount in exchange for the right to reprint these plain-English instructions and distribute them to other network customers.

The second compromise has to do with phasing in the LAN implementation. Even though each manager would like his or her department to be the first phase of the installation, the intradepartmental functions this would serve have already been identified by the committee as not-high-priority. Linking Nyack workstations with the typesetting center is a high priority, but it's also an expensive first step. Weighing the potential return against the initial cost of other possible first phases, the committee identified the mini-to-managers links as the optimum first phase of the installation. Since this stretch of the network will also serve as the main trunk of the eventual tree topology, installing it first will also facilitate expansion later.

End of case study (Part V)

SECTION FOUR:
AFTER THE LAN, WHAT?

Chapter 10

Linking LANs

One thing you may have noticed about the computer and communications technology that's made its way into your business in the last decade or so: It's addictive. However intimidated your managers were by the first microcomputers plunked on their desks, however resistant your clerical people were to learning word processing, however hostile your sales clerks were to their first point-of-purchase terminals—they wonder now how they ever did their jobs without these tools. That doesn't mean they'll rush to welcome every new electronic tool that comes down the pike, but don't try to take away the devices they've come to depend on. Offer them a new technological resource they can master in under 30 minutes, without having to unlearn what they already know, and odds are they'll embrace that tool, too.

In company after company, as users get past the computerphobic threshold, they begin to invent new ways to use their terminals and workstations. Mainly, they do this to save themselves time and aggravation; a few old-fashioned types do it to save their employers money. Once you've begun to implement your local area network, the same kind of momentum will obtain. Wouldn't it be great, your networked staffers will say, if *that* work group could be hooked into "our" network and we could get our weekly, monthly, and quarterly reports from them electronically? This is how enterprise-wide networks grow from relatively small, self-contained LANs.

Wherever the inspiration to expand comes from in your company, you'll be glad to know that such growth is possible, even if it wasn't part of your original master plan. Naturally, it's easier if you did anticipate it, and if you did think about *how* you'll want your network to expand.

HANGING MORE STATIONS ON THE LAN

As noted in Section Two, it's possible to add stations to virtually any local area network, up to limits dictated primarily by the medium. Doing so does result in some trade-offs:

- Adding stations will slow down network response time, eventually creating delays that users may find intolerable.

- Adding stations to a ring-topology network requires disabling the system during installation.

- Adding stations to a network with a contention access method like CSMA/CD will eventually cause traffic problems that could shut some stations completely out of the network.

- Adding stations to a slotted or polled access network will reduce the window of opportunity each station has to use the network.

How do you avoid bumping into these ceilings if you want to keep expanding your network?

A NETWORK OF NETWORKS

The answer is almost too obvious: You build a bridge from one small, discrete network to another. The routine within your accounting department needn't share network capacity with data that only your production department needs, but when the two work groups do need to pass data back and forth—or when even one user in accounting needs data from production – there's an electronic path for that data. It's comparable to the stretch of U.S. 95 that links the Baltimore and Washington Beltways.

There's more than one way to create such a LAN-to-LAN path: In ascending order of sophistication, *routers*, *bridges*, and *gateways* are the official names for these linkages.

Routers

Though the technology of networks is still quite young, the extra capabilities of bridges and gateways have already made routers almost passé. Routers do exactly what their name implies: They direct traffic from one network segment to another, performing no translation or protocol conversion in the process; this means they can link *only* identical networks. If identical networks are what you need to link, a *router* will do the trick for you.

Bridges

But what if you've installed one kind of network in accounting and another in production? Then you need a network-to-network link that can perform some protocol conversion – a link that not only pushes data from one network to another, but rearranges it into the shape it needs to assume to travel on the second network. For this task you need a *bridge* (Fig. 10-1).

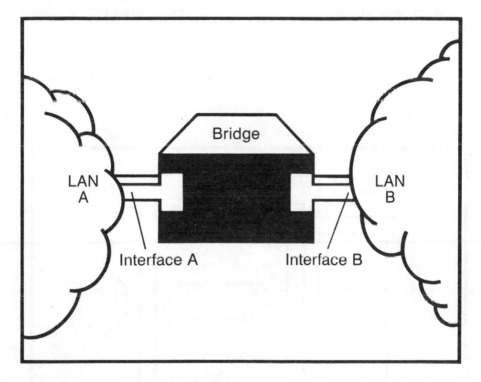

Fig. 10-1. Bridge.

Gateways

Protocol conversion notwithstanding, a bridge still only moves raw data from one network station to another. Suppose you want that raw data—last month's sales figures, for example—from one network station to be received at another network station in a different format—say, plugged into the appropriate spots on the receiving station's spreadsheet. For that you need routing, plus protocol conversion, plus translation, plus application support functions. For that you need a *gateway* (Fig. 10-2).

Clearly, a gateway is the most versatile of the LAN-to-LAN links, and the one you're most likely to want if, like most organizations today, you've acquired a mix and match assortment of computer equipment and software from several different suppliers. But any of these links may meet your particular needs, and you can (and should) consider them when you think about the ways you might want to expand your LAN to serve a larger number of in-house users.

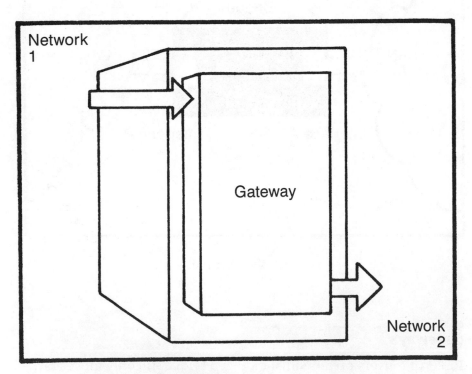

Fig. 10-2. Gateway.

REACHING OUT

Suppose you want to expand the reach of your LAN to include out-lying offices or customers or suppliers or the world in general? Can you turn your LAN into the first paving stone of such an enlarged information pathway? Indeed you can, up to a point—and that point is improving literally every day.

One of the defining characteristics of all LANs cited in Chapter 1 is that a local area network is *local*. Its scope is generally limited to about two miles. To reach beyond that local limit and shake hands with another LAN or a larger network, you'll want a gateway.

Gateways can be used not only to connect one LAN to another, but also to connect LANs to:

- Wide area networks (WANs).

- Private data communications networks, such as high-speed data lines ("T1 links") leased exclusively to your company by the local telephone operating company or private networks operated by information services companies and subscription databases.

- Long-distance telephone networks, including satellite services.

- The public switched network (aka "the telephone system"). In the next decade the public network will be evolving into the Integrated Services Digital Network (ISDN). The LAN-ISDN connection will take on even greater significance than today's LAN-to-public-network connections.

For the time being, *only* gateways can perform all the protocol conversion and translation required to move data from a LAN to one of these larger networks. The conversion and translation required may be different for each of the larger networks you want to access, so you may need a different gateway for each one. This will certainly be true if you want to access both an all-digital T1 link and the still-mostly-analog public switched network.

THE LIMITS OF GATEWAYS

If LAN outreach is part of your overall networking plan, you'll want to ask the vendors you talk with about the gateways available with the networks they offer. The tricky thing about today's gateways is that they are A-to-B connections only. To link A to B and C and D, you need three gateways. To link A to B to C to D, you need six (A-B, A-C, A-D, B-C, B-D, and C-D). To interconnect ten discrete networks, you need 45 gateways (Fig. 10-3)!

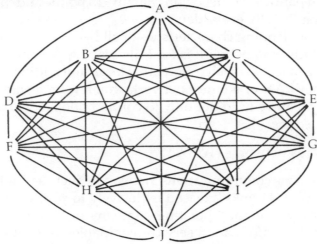

Fig. 10-3. 45 Gateways.

This is a complicated solution, to be sure. Even more dismaying is the fact that not all the gateways that are *theoretically* possible actually exist. It's critical, therefore, that you consider what gateways you might want or need, and ask the vendors you talk with if they can meet those needs. If your long-range plans call for tying Vendor X's LAN to Vendor Y's, your vendor had better have created the gateway you need to do just that. This is a particularly important consideration if you want or need to network highly specialized equipment from a small, specialized vendor. Network vendors have created gateways to link nearly all the most popular brands of computer equipment; the law of supply and demand applies. And some network vendors have made a specialty of creating links for unique equipment. If you have such equipment, you need to know how (or if) your vendor can tie it into your network.

The good news is that less-complicated solutions are on the way. The better news is that you don't have to wait for them: the LAN you install now can be building block number one in the network that connects you with information resources around the world—if you *plan* for that kind of connection. We'll look at some of the work being done in this area, and ways you can plan today for tomorrow's global network, in Chapter 11.

CASE STUDY—Part VI

Fifteen months after implementation began, Electromedia has phased in one local area network at the Nyack site and another at West Nyack. Each of the headquarters departments and divisions has its own mini-network joined to the main trunk of the LAN, and the efficiencies achieved through shared, timely information are already showing up in Electromedia's bottom line.

The real 'great leap forward,' however, depends on creating the link between Nyack and West Nyack, for that connection will optimize the typesetting resources available to all the Nyack divisions. And Sly Fachs is still eager to get the Rochester outpost tied into the company's network.

Although the West Nyack site is near enough to be part of Electromedia's local area network, securing rights of way to run fiber-optic cable from one site to the other makes that impracticable. One option is to link West Nyack's LAN via a gateway to a private leased telephone line linked by another gateway to the Nyack LAN. Sly Fachs and an executive committee of the original LAN committee have looked into this, and Sly has balked at the cost of a private leased line.

So the network vendor asked if the committee had considered a "virtual private network" — actually a shared public facility, but software-controlled in such a way that it functions as a "private" network for each of the users sharing it. Sharing lowers the cost without limiting any user's access to the network, the vendor explained. A gateway connects each LAN to the virtual private network, in the same way it would connect each LAN to a private leased line.

What's more, the same kind of gateway-plus-virtual-private-network connection can be used—at last!—to tie the software facility in Rochester to the Nyack LAN, and to the West Nyack LAN, as well. Thus, just eighteen months after implementation began, Electromedia Inc. has finally become a fully networked organization.

End of case study (Part VI)

Chapter 11

Universal Connectivity

It didn't take long for computer users to discover that these labor-saving devices could save them even more labor—not to mention time, aggravation and money—if they could communicate with each other. The desire for computer-to-computer communication soon became a necessity, and that necessity was the mother of computer networking. "We want to link the computers you sold us," customers told their vendors, and the vendors brought forth networks.

It wasn't an easy task. Few of those computers had been designed to communicate with each other in the first place, so the vendors sometimes had to create bridges and even gateways to link one piece of their equipment with another. Linking Vendor A's equipment with Vendor B's or C's was a problem nobody even wanted to address—until users began demanding such links. By then, each vendor had developed its own unique communications "architecture" and the creation of "multi-vendor networks" grew more difficult even as it became more necessary.

IS THIS ANY WAY TO RUN A RAILROAD?

The existence of all those proprietary network architectures has been compared to having every county in the U.S. set its own unique requirements for railroad tracks: Moving cargo (data) from one railroad system (network) to another required a lot of manual labor that would have been unnecessary if all the railroads (networks) had been built the same way in the first place. And why *shouldn't* there be one standard data network, the way there's one standard telephone network anyone in the world can use?

Of course, one reason so many different data network architectures existed was that they were created by competing companies—unlike the telephone network, which was created mostly by national monopolies. Now the problem was to decide *which* existing network architecture should be adopted by everyone else as the standard. That's a little like asking the United Nations to agree on which human language ought to be designated the world's only official language. Just try to achieve unanimity on a question like that. Just try to make it work without unanimity.

A De Facto Standard

For a time, it seemed that IBM's Systems Network Architecture (SNA) would become the de facto industry standard—chiefly because IBM's installed base was so large. Not surprisingly, IBM was the only company that was wildly enthusiastic about this prospect. In addition to the competitive edge such a standard would give to IBM, some experts objected that SNA fell short, in various ways, of being *the best possible* network architecture. Finally, if the industry's vendors were to agree on an existing architecture as the standard, the company that created that architecture would need to make it "open"—share the details of how it works with all other vendors—and there are good reasons for IBM (or any vendor) not to want to give away such information. For competitive as well as technical reasons, the industry was just not going to embrace an existing architecture as the network standard.

The alternative was to create independently a new network standard—or set of standards—that didn't match anything already in existence and to publish the specifications so that any vendor who chose to could build its computers and communications equipment and services to access the resulting "open" network, the way telephone equipment manufacturers build their units to access the existing telephone network. The work of defining "open network standards" has been going on for more than a decade in existing standards organizations, including the International Organization for Standardization (ISO), the International Telephone and Telegraph Consultative Committee (CCITT), the National Bureau of Standards (NBS), the American National Standards Institute (ANSI), the International Telecommunications Union (ITU), and the Institute of Electrical and Electronic Engineers (IEEE). In particular, ISO has developed a seven-layer Open Systems Interconnection (OSI) Reference Model (Fig. 11-1) as a framework for the definition of pertinent standards.

The OSI model is not, in itself, a standard, but an organizational framework that encompasses dozens of standards at each layer. The individual standards are developed, defined and promulgated by ISO-affiliated standards bodies, who either start from scratch to formulate a standard protocol or adapt vendor-developed protocols. It

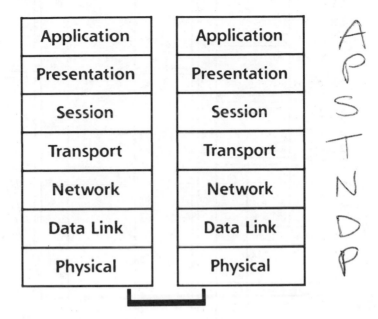

Fig. 11-1. ISO's seven-layer model.

is a time-consuming and complicated process—and an evolutionary one. New standards will continue to be needed and developed as human needs to communicate continue to evolve, and as communications technology continues to advance.

In the area of LANs, for example, standards promulgated by the IEEE Committee 802 (and labeled 802.3, 802.4, etc.) have now been embraced by the International Organization for Standardization and inserted into layers one and two of the seven-layer OSI model as ISO Draft International Standards (DIS) 8802.3, 8802.4, etc. (Fig. 11-2)

LAN standards are only part of the much bigger open communications picture. They are an important issue for organizations like yours, seeking to maximize the utility and benefits of corporate information now, but their role as a building block in the open *global* network of tomorrow is the real reason they demand attention.

MAKING IT HAPPEN

Much work remains to be done before the universal open network everybody wants becomes a reality. But much *is* being done even

Fig. 11-2. The use of the OSI model for global communications. (From "World Communications," Siemens AG, Munich, Germany. Reprinted with permission.)

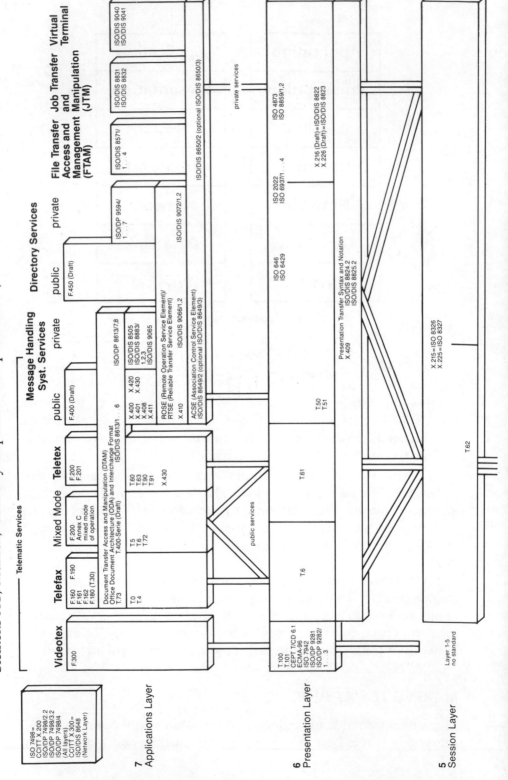

4 Transport Layer

Connection-mode service (COS)

ISO 8072=X.214
ISO 8073=X.224

T.70

class 0-4 class 4

T-COS, C-COS, P-COS, I-COS, L-COS   C-CLS, L-CLS

Connectionless application, at present no International Standard

ISO 8072
ISO/DIS 8602

Service type conversion is:
= not necessary or mandatory
= provided in the Network Layer
= opt. prov. in the Transport Layer

C-CLS, P-CLS, L-CLS
T-COS, C-COS, P-COS, I-COS, L-COS

Connectionless-mode service (CLS)

---

3 Network Layer

T — COS
ISO/DIS 8348 =rev. X.213
ISO 3166
ISO 6523
E.163
V.25 bis
T.70/X.25

C — COS
ISO/DIS 8348 =rev. X.213
ISO 3166
ISO 6523
ISO/DIS 8208
ISO 8878
ISO/DIS 8880/1
ISO/DIS 8880/2
T.70/X.21 X.121

C — CLS
ISO/DIS 8348
ISO 3166
ISO 6523
ISO/DIS 8473
ISO/DIS 8880/1
ISO/DIS 8880/3
X.21
X.21

P — COS
ISO/DIS 8348 =rev. X.213
ISO 3166
ISO 6523
ISO/DIS 8208
ISO 8878
ISO/DIS 8880/1
ISO/DIS 8880/2
T.70/X.25
X.25 X.121 X.244

P — CLS
ISO/DIS 8348
ISO 3166
ISO 6523
ISO/DIS 8208
ISO/DIS 8473
ISO/DIS 8880/1
ISO/DIS 8880/3
ISO/DP 9068
X.25 X.121 X.244

I — COS
ISO/DIS 8348=rev. X.213
ISO 3166   I.330=E.164
ISO 6523   I.331=E.164

D-Channel
I.450=Q.931
I.451=Q.931

I.461=X.30
I.462=X.31
I.463=V.110

B-Channel
I.462=X.31

L — COS
ISO/DIS 8348 =rev. X.213
ISO 6523
ISO/DIS 8208
ISO 8878
ISO/DIS 8880/1
ISO/DIS 8880/2
ISO/DIS 8881/1.2
ISO/DIS 8881/2.2

L — CLS
ISO/DIS 8348
ISO 3166
ISO 6523
ISO/DIS 8473
ISO/DIS 8880/1
ISO/DIS 8880/2
ISO/DIS 8880/3
ISO/DIS 8881/1.2
ISO/DIS 8881/2.2

---

2 Data Link Layer

T
ISO/DIS 8885
T.71

ISO 3309
ISO 4335
ISO/DIS 7478
ISO 7809
ISO 8471

C
ISO/DIS 8885.2

ISO 3309
ISO 4335
ISO/DIS 7478
ISO 7809
ISO 8471
T.70/X.75
X.21 (SYN SYN)

P
ISO/DIS 8885.2

ISO 7776
ISO/DIS 8880/1
ISO/DIS 8880/2
ISO/DIS 8880/3
T.70/X.25
X.25

I
D-Channel
I.440=Q.920
I.441=Q.921

ISO 3309
ISO 4335
ISO 7809

D- and E-Channel
ISO 8471
ISO/DIS 7478
ISO/DIS 8885
I.462=X.31

L
ISO/DIS 8802/2.2
ISO/DIS 8802/3
ISO/DIS 8802/4
ISO/DIS 8802/5
ISO/DIS 8802/7

ISO/DIS 8802/2.2
ISO/DIS 8802/4
ISO/DIS 8802/5
ISO/DIS 8802/7

---

1 Physical Layer

PSTN
ISO/DIS 2110, V.28, V.24, V.
ISO/DIS 2110, V.31 od V.31bis: V.24
ISO/DIS 2593, X.35, V.35: V.24
or: V.11: V.24, V.36, V.10
ISO/DIS 4902 V.10

CSPDN
ISO/DIS 2110, X.21bis: V.24, V.28
ISO/DIS 2593, X.21bis: V.24, V.35
ISO/DIS 4902, X.21bis: V.24, V.36, X.26, X.27
ISO/DIS 4903 X.21: X.22, X.24, X.26, X.27

ISDN
ISO/DIS 8877
I.430
I.431
I.460
I.461=X.30
I.462=X.31
I.463=V.110
I.464

LAN
ISO/DIS 8802/2.2
ISO/DIS 8802/4
ISO/DIS 8802/5
ISO/DIS 8802/7

PSTN Public Switched Telephone Network
CSPDN Circuit Switched Public Data Network
PSPDN Packet Switched Public Data Network
ISDN Integrated Services Digital Network
LAN Local Area Network

Note: Addenda are not explicitly shown, because they are effective implicitly in conjunction with the base International Standards.

now, and some experts predict the existence of "an OSI world" within the next decade. Two important elements in this evolution are the Integrated Services Digital Network (ISDN) and organizations like the Corporation for Open Systems (COS).

Integrated Services Digital Network

ISDN is, in a way, an outgrowth of the universal telephone network. What's envisioned is the eventual replacement of that mostly analog, mostly voice-traffic global network with one all-digital network that will carry data, video, facsimile and other digitized traffic, as well as digitized voice transmission. Picture a wall outlet you can plug any computer-based device into, the way you can plug any telephone into a phone jack today, to send or receive information around the world: That is the goal of ISDN.

ISDN will transform the public network into a supercarrier of all kinds of information and services—some of them still undreamed-of. Where today's single-channel telephone line allows you to transmit data at a rate of 9.6 kilobits per second, the first incarnation ("basic access") of ISDN will offer you two 64 Kbps "bearer" (B) channels for voice and data, plus a 16 Kbps "data" (D) channel for packetized data and network control. As the public network evolves toward all-fiberoptic installation, "primary access" ISDN will offer *twenty-three* 64 Kbps B channels, plus the 16 Kbps D channel. That is more than 150 times the capacity of a single telephone line in today's network.

Work on ISDN has progressed somewhat more rapidly in other countries than in the U.S. One reason is that telecommunications is a government-controlled (often government-owned) monopoly in other countries, and the absence of competitive pressures makes it easier to steer a straight course toward the establishment of a network like ISDN. Another reason is that user demand for such an international network is greater in Europe and the Far East, where commerce across national boundaries is more common and more vital to the local economy.

Nevertheless, the major communications carriers in the U.S. – including the regional and national companies that used to be AT&T, as well as newer long-distance carriers—are all pursuing ISDN goals and timetables. Field trials of ISDN have been conducted by U.S. companies here and abroad, and deployment is expected to begin by 1991. It is worth asking the LAN vendors you talk with whether ISDN has been factored into their plans.

ISDN standards are based on the OSI Reference Model; ISDN's planners intend for it to be the uniform railroad track that carries data from one device or network to another. The other part of the "universal connectivity" equation is building devices that can tap into that network—building railroad cars that fit that track. That is where the Corporation for Open Systems and its overseas counterparts, Europe's Standards Promotion Application Group (SPAG) and Japan's Conference Promoting OSI (POSI), come in.

Corporation for Open Systems

COS is a coalition of computer and communications vendors and users with the shared goal of seeing communications standards *implemented*—that is, actually built into networkable products. Founded in 1985 by 20 major computer and communications vendors, the organization had tripled in size two years later and now includes nearly every major vendor in the information industry.

COS founders agreed that there was no need for yet another standards-setting organization; they embraced the OSI Reference Model at the outset, and have concentrated on selecting specific standards that fit the model, disseminating information about those standards, and testing equipment to verify that it does incorporate the selected standards and therefore will interoperate with other COS-certified equipment. Some observers expect the COS certification sticker, when it begins to appear on equipment next year, to be comparable to the Underwriter's Laboratory label on electrical equipment. Others compare it to the "Beta" and "VHS" identification on video cassettes. Equipment buyers seeking to build or enlarge OSI networks will look to the sticker for assurance that they've chosen compatible, connectable equipment.

Even before COS testing and certification began, the organization gave momentum to OSI and related concepts like ISDN. Education and information-dissemination efforts have spurred users to think futuristically about how a product or service will mesh with the open, global network of tomorrow. It is an especially pertinent question for LAN shoppers to ask. Once that universal, open network is in place, an isolated LAN will seem as anachronistic as stand-alone computers do now.

CASE STUDY — Part VII

It's the year 2000, and Electromedia Inc. has been fully networked for more than a decade, enjoying all the advantages of sharing resources that it had anticipated when Sly Fachs first convened his LAN committee. But networking has done more for Electromedia than cut costs and improve efficiency. It has actually created new profit opportunities — new lines of business even Sly Fachs hadn't envisioned when he first began looking at LANs.

It almost happened by accident, and yet it seemed such a logical, natural extension of the intra-company sharing the network made possible: With the LAN in place, the editors and market researchers and sales staffs of Electromedia's divisions were able to share information about developments in the industries Electromedia publications served. That fact alone added value to the publications and to their mailing lists, and the direct-marketing division could use that information, now part of a de facto company-wide database.

Then one day a particularly astute marketeer suggested to Sly that the information in the database was itself marketable, and Electromedia got into the marketing consultancy business. A few months later, one of Electromedia's technical wizards tried again to convince Sly that electronic publishing — public databases — was the direction the media industry was heading, and this time Sly listened, because the network made it an opportunity that could be seized for a relatively small incremental cost.

The worldwide implementation of multi-service all-digital networks created still more opportunities for Electromedia to market the information it gathered, in an ever-wider array of formats — not just words, but pictures, even moving pictures — and made it the world's leading purveyor of information about information.

End of case study (Part VII)

Appendix I

Local Area Network Manufacturers

Altos Computer Systems
2641 Orchard Parkway
San Jose, CA 95134
(800) ALTOS-US

Worknet is a general data processing/office automation LAN connecting up to 30 Altos multi-user microcomputer systems. The physical length of the network is 1,500 meters. A user can also link an IBM Personal Computer to this network with the installation of an additional product called PCpath. Altos Computer Systems sells through distributors. Contact the company for the name of your nearest dealer.

Apple Computer
20525 Mariani Avenue
Cupertino, CA 95104
(408) 996-1010

LocalTalk connects up to 32 Apple personal computers on a LAN that is up to 300 meters (1,000 feet) long. LocalTalk, introduced in 1985, is available through a dealer network. Contact your local computer dealer or call (800) 538-9696.

Applitek Corporation
107 Audubon Road
Wakefield, MA 01880
(617) 246-4500

Applitek's UniLan is a general-purpose, large intrafacility LAN, suitable for installation at campus sites. It provides support for many different types of asynchronous and synchronous devices, along with gateways to allow communication with IBM, Sperry, DEC, Burroughs, and UNIX Multibus hosts. Media used include baseband coaxial, broadband coaxial, and fiberoptic cable. UniLan was introduced in 1983 and now is installed at 132 sites. Applitek markets through direct sales and prime contractors.

AST Research Inc.
2121 Alton Avenue
Irvine, CA 92714
(714) 863-1333

AST research, Inc. currently offers three local area network solutions to attach IBM PCs, XTs, ATs, or 100% compatibles together and allow them to share data and peripherals.

The AST-StarPort is designed as a cost-effective, entry-level solution for attaching a maximum of ten PCs within a segment distance of 400 feet in a bus-type topology utilizing twisted pair technology. Data transmission rate of the StarPort is 1 Mbps.

The Resource Sharing Network (RSN) adapter which uses AST's own proprietary 5 Mbps physical link to RG-59 cable, was designed to fit into the moderately sized local work group.

The Ethernet adapter, AST's newest LAN attachment product, is fully IEEE 802.3 compatible. This adapter is AST's answer to the larger sized LAN work group. Targeted toward those requiring heavy LAN usage, the AST-Ethernet adapter is currently the fastest 10 Mbps adapter on the market.

All adapters are certified to operate under Novell's NetWare 2.0a and 2.1, Corvus's PC-NOS, and Tapestry.

Astrocom, Inc.
120 West Plato Boulevard
Saint Paul, MN 55107
(612) 227-8651

XLAN is a general connectivity LAN accommodating an ASCII, asynchronous or RS-232 device from vendors such as Apple, IBM, and Hewlett-Packard. XLAN supports up to 192 devices, which could include personal computers, terminals, minicomputers, and modems. XLAN has been on the market since 1983, and has been installed in 347 sites across the country. The product is sold through a network of distributors nationwide. Contact Astrocom, Inc. for the name of a distributor near you.

AT&T Information Systems
(800) 247-1212

AT&T's Information Systems Network (ISN) supports networks situated in buildings and campus sites, as well as long-distance, interpremises networks. ISN provides support for RS-232C devices, SDLC bisynchronous terminals, IBM 3278 asynchronous terminals and IBM 3274 cluster controllers. It links devices adhering to the Ethernet 802.3 specification and supports AT&T and DEC VAX processors running UNIX operating systems. ISN can be linked to AT&T PBX switches such as the System 75, System 85, Dimension 3000 or 6000. ISN interfaces with AT&T's Starlan Network, which serves the AT&T PC 6300, Unix PC, 3B2 processor and the IBM Personal Computer. ISN makes use of the Premises Distribution System (PDS), developed by AT&T, which carries voice, data, graphics, video, and sensor signals over a combination of fiberoptic and twisted-pair wire. ISN has been available since 1984. For further information call your local AT&T account executive.

Concord Communications Inc.
753 Forest Street
Marlboro, MA 01752
(617) 460-4646

Token/Net (introduced in 1983) and MAPware (introduced in 1986) are two LAN products focused toward factory automation applications. Token/Net supports any RS-232 or RS-449 device, including the DEC VAX line and personal computers. MAPware requires a specific bus structure — any type of computer with a Multibus or PCbus, including the IBM PC industrial computer will run on the network. Both the Token/Net and MAPware will handle 1,000 plus nodes. The products are sold directly by Concord Communications.

Control Data Corporation
Computer Systems Marketing
P.O. Box O, IIQW09G
Minneapolis, MN 55440
(612) 853-7696

Loosely Coupled Network (LCN) permits multiple Control Data computer systems, as well as other large-scale processors from various vendors, to communicate using a set of common data trunks.

LCN is intended for high-volume, processor-to-processor file transfers. The data transfer rate is 50 Mbps. As many as 100 systems can be connected to a four-trunk network. The end-to-end maximum cable length is 3,000 feet of coaxial cable. Control Data Distributed Communications (CDCNET) enables users to connect host computers, terminals, and workstations from a variety of different vendors into data processing networks. The networks connected with CDCNET can range in size from a single local area network to multiple local area networks linked together by communications lines that span long distances. At least one Cyber Computer system must reside on the network.

Corvus Systems Inc.
160 Great Oaks Boulevard
San Jose, CA 95119
(408) 281-4100

Omninet is a personal computer-based LAN that connects up to 64 personal computers. Omninet lets the user create an all-Macintosh, all-DEC, all-Apple, all-IBM network, and others. In addition, any combination of the above systems is available. Omninet is available through distributors and dealers. For information call (800) 4-CORVUS.

Data General
4400 Computer Drive
Westboro, MA 01580
(617) 366-8911

Data General's Ethernet local area network adheres to the IEEE 802.3 specifications and supports Ethernet across its entire product line. Its LAN can be used for general data processing applications and engineering environments — or both on the same LAN. Data General's LAN will support any equipment, including DEC, Wang, IBM, or any other equipment that supports the Ethernet 802.3 standard. The company markets through direct sales, as well as value-added resellers.

David Systems
701 E. Evelyn Avenue
Sunnyvale, CA 94086
(408) 720-8000, Ext. 290

David Information Manager supports any Ethernet-compatible hardware and software as well as any RS-232 equipment. It will handle small networks of 24 users, with up to 3,000 Ethernet connections, or 6,000 RS-232 connections. David Information Manager was introduced in October of 1984, and is sold directly by David Systems, as well as through Ameritech, one of the regional Bell operating companies. David Systems has installed over 450 networks.

Digital Equipment Corporation
146 Main Street
Maynard, MA 01754-2571
(617) 897-5111

DECnet is hardware, software, and distributed applications that allow up to 64,000 computer systems to form a network in a LAN or WAN environment. DECnet is based on the seven layer Open Systems Interconnect (OSI) reference model. Ethernet is DECnet's LAN of choice and DECnet connects LANs together using telephone lines, fiber optics, microwave, and satellite links. Digital supports Ethernet over broadband, and three forms of baseband, including traditional thin-wire coax, and unshielded twisted pair cable.

DECnet is supported by Digital's time sharing, personal, and workstation computers. DECnet is also available for the IBM Personal computers through Digital, and for other vendors' computers such as Apple Computers through independent software vendors.

Contact your local sales office or call the company directly for information.

DSC Communications Corp.
3101 Scott Boulevard
Santa Clara, CA 95054
(408) 727-3101

DSC's personal computer-based LANs, called the Plan Series, connect IBM Personal Computers with shared system elements ranging from printers to disk storage to mainframes. The Plan Series are positioned for large institutional end users. DSC sells through a direct sales force and third-party distribution. It provides all the elements, hardware and software, to achieve a fully integrated personal-computer network. The Plan Series' network configurations can be either token ring or ARCnet. Interconnections can be either standard coaxial cable, shielded twisted-pair, or fiberoptic. For more information call the company directly.

Excelan Inc.
2180 Fortune Drive
San Jose, CA 95131
(408) 434-2300
(800) EXCELAN

Excelan Open Systems (EXOS) link dissimilar computer systems over an Ethernet LAN. Excelan's high-speed, intelligent controllers running TCP/IP allow IBM PCs to talk to one another and also to VAXs, MicroVAX IIs, Sun workstations, and UNIX supermicros. IBM PCs will communicate with minis and mainframes running UNIX, VMS, and Xenix operating systems. Excelan networks are used predominantly in such applications as design, office, government and factory automation. Contact the company directly for information.

Gateway Communications, Inc.
2941 Alton Avenue
Irvine, CA 92714
(714) 553-1555; outside California, (800) 367-6555

Gateway Communications manufactures local area (LAN) and wide area networking (WAN) products for IBM PCs/ATs/XTs/PS/2-30s, and compatibles. Gateway's modular family of low-cost, systems-oriented solutions helps companies interconnect PCs to share resources and information in a workgroup environment.

LAN topologies include: the proprietary G/Net LAN; the IEEE 802.3 G/Ethernet LAN; and the IEEE 802.5 G/Token-Ring LAN. WAN products include: G/Remote Bridge, the first high-speed remote bridge for any NetWare-based LAN; G/SNA Gateway, a high-speed multi-user LAN-to-mainframe gateway; and G/X25 Gateway, a high-speed, multi-user LAN-to-remote host gateway.

Honeywell Bull Inc.
P.O. Box 8000
Mail Station B39
Phoenix, AZ 85066
(800) 238-5111

Honeywell Bull offers a line of Ethernet-type LANs from Bridge Communications. This is a general-purpose LAN with a data transfer rate of 10 Mbps. Honeywell Bull also offers a personal computer-based network from 3-Com that will interconnect with the Ethernet-type LAN. Honeywell Bull offers GM/MAP, a broadband token bus LAN from Concord Communications Inc. Contact your local Honeywell Bull office for more information.

IBM Corporation
Old Orchard Road
Armonk, NY 10504

The IBM PC Network is a broadband LAN designed to link IBM Personal Computers. The IBM PS2 family of computers plus earlier versions including the IBM PC, PC XT, Portable PC, and PC AT can be connected to the network using CATV coaxial cable. As many as 72 IBM PCs can be connected in the network, at a maximum distance of 1,000 feet from a network translator unit.

The IBM Token Ring Network is a high-speed communications network for information processing equipment that uses the IBM cabling system, including Type-3 telephone media and a token ring protocol. The network permits data transmission at 4 Mbps. It supports the attachment of most IBM computer systems, the direct attachment of IBM communications controllers, and the interconnection via gateways of the Series/1 and System/36 computers. Primary users can be an enterprise or a department within an organization. Contact your local IBM representative.

InteCom Inc.
601 InteCom Drive
Allen, TX 75002
(214) 727-9141
(800) INTE-800

LANmark is a voice / data LAN that works with InteCom's Integrated Business Exchange (IBX) and supports the Ethernet 802.3 and 3270 standard. It utilizes a building's existing twisted-pair wire, instead of the more expensive coaxial cable typically used with Ethernet LANs. InteCom markets directly and through distributors, as well as through TTS in Canada and Italtel in Italy.

Magnolia Microsystems Inc.
2818 Thorndyke Avenue West
Seattle, WA 98199
(206) 285-7266

MagNet links IBM Personal Computers and compatibles, IEEE 696 bus computers and Z-80 microcomputers for general office applications. MagNet will accommodate up to 64 nodes and uses shielded twisted-pair wiring as its medium. The maximum length without repeaters is 2,000 feet. MagNet is sold directly by Magnolia Microsystems.

McDonnell Douglas Computer Systems Co.
17481 Red Hill Avenue
P.O. Box 19501
Irvine, CA 92713
(714) 250-1000

The Reality Operating System—Extended has local and wide-area networking facilities designed into the operating software. These extremely powerful networking facilities are independent of the physical transport service used.

For local connection of systems, McDonnell Douglas provides an Ethernet compatible local area network connection. The S-LAN, as it is known, uses IEEE 802.3 electrical interfacing, making it hardware compatible with standard Ethernet. The physical network links and hardware controllers used by ROS-E adhere to Open

System Interconnection standards of the International Standards Organization for present and future compatibility with other systems and worldwide networks.

Networking Facilities: Remote logon is the facility that allows any user with authorization to log on to any connected network member. In most cases the user need not even know that there is a network linkage involved.

Port contention is the ability to connect more terminals to a system than there are tasks to handle them. Port connection will notify the user of the lack of resource and give them a place in a queue for later processing.

Applications may start other remote or local tasks, making the structure of complex applications systems easier to manage. Any task may request the start of a server task in the local or any remote system. Dialog then occurs directly between the programs until one or the other terminates the relationship. This method of defining related programs facilitates the use of intelligent dialog that minimizes the use of network resource and makes practical distributed applications a reality.

Network Systems Corp.
7600 Boone Avenue North
Minneapolis, MN 55428
(612) 424-4888

HYPERbus, a standard terminal network, connects RS-232 equipment and 3270-type terminals to various mainframes. It is a 10 Mbps network utilizing coaxial cable. HYPERchannel-10 allows up to 30 different vendors' equipment to interconnect at a transmission speed of 10 Mbps. Micro-to-mainframe, terminal-to-mainframe, and mainframe-to-mainframe communication are supported. Applications include file transfer and direct memory access. HYPERchannel-10 utilizes twisted-pair and fiberoptic cable and is compatible with HYPERbus networks.

HYPERchannel-50, designed for mainframe-to-mainframe file transfers, transmits at speeds of 50 to 200 Mbps using from one to four trunks. Forty different operating systems are supported. HYPERchannel-50 uses coaxial, fiberoptic, T1 or T3 telephone lines, microwave, or satellite networks. Remote Device System (RDS) is a point-to-point channel extension network that allows high-speed peripherals to be located away from computer centers. Transmission speed is 50 Mbps.

HYPERchannel-DX, a new product line, provides high-performance integration of different networks, such as Ethernet, HYPERchannel, Token Ring, and FDDI. A single HYPERchannel-DX unit can route datagrams from different protocols (such as NETEX, TCP/IP, OSI, or DECnet) throughout a network of networks. Its design permits various configurations and up to 16 concurrent "conversations." The HYPERchannel-DX memory can move data at up to 400 Mbps. Therefore, host interfaces such as the full-duplex 100 Mbps Cray channel can operate without performance degradation.

Novell Inc.
Communications Department
122 East 1700 South
Provo, UT 84601
(800) 453-1267

NETWare is an IBM PC AT or compatible local area network that allows for sharing of data and peripherals. IBM, SNA, and DEC gateways are available. Novell sells mainly through value-added resellers and distributors. Call the 800 number for more information.

Prime Computer Inc.
Prime Park
Natick, MA 01760
(617) 655-8000

The RINGNET token-passing, general purpose LAN uses proprietary protocols to route data. The PRIMENET distributed networking facility offers wide area capabilities over X.25 for connection of Prime and non-Prime systems. Prime/SNA communications software, Prime's implementation of IBM's Systems Network Architecture, allows Prime systems to coexist with IBM SNA computer networks. Prime/SNA operates with PRIMENET software to extend Prime/SNA capabilities across multiple Prime systems and networks. PRIMIX operating system software is a programming environment based on the AT&T UNIX System V operating system that allows Prime systems to communicate with other systems running UNIX. PRIMELINK communications software provides a link between IBM-compatible and Apple Macintosh personal computers and any host 50 Series system. Prime Computer markets through direct sales and authorized resellers.

PRIME and the Prime logo are registered trademarks of Prime Computer, Inc., Natick, MA.

RINGNET, Prime/SNA, PRIMENET, PRIMIX, PRIMELINK and 50 Series are trademarks of Prime Computer, Inc., Natick, MA.

Proteon Inc.
Two Technology Drive
Westborough, MA 01581
(617) 898-2800

Proteon has three token ring networks: ProNET-4, a 4 Mbps IEEE 802.5 and IBM compatible network, which supports IBM Personal Computers, PC ATs and Multibus-based and VME bus-based workstations.

The company offers a variety of IEEE 802.5 compatible Wire Centers (MAUs) and both copper and fiber optic repeaters. Also offered is TokenVIEW-4 Network Management, which proactively manages IEEE 802.5 and IBM compatible token-ring networks. TokenVIEW-4 assists in network fault management and configuration management.

ProNET-10, a 10 Mbps network supporting Unibus for VAX VMS and Q-bus for MicroVAX, also has interfaces for Multibus and VME-bus, Sun workstations, and IBM PCs, XTs, ATs, and PS/2s. ProNET-10 supports up to 255 hosts over distances of 10 miles. Network management is available with TokenVIEW-10 to maximize ProNET-10 availability. It can be used for general business applications and high performance applications like CAD/CAM.

ProNET-80 operates at 80 Mbps, linking mainframes, minicomputers, PCs and workstations. Because of its high speed, it is ideal for real-time applications, file transfers, and computer-aided design. Equipment supported includes VAX VMS, PDP-11, LSI-11, Micro PDP-11. It also has interfaces for Multibus, Sun II workstations, Masscomp, and Intel workstations, the IBM PC AT, and VME bus interfaces. ProNET-80 supports up to 240 hosts. It uses fiberoptic cable as well as IBM cabling system. Proteon has more than 1,000 network installations. It sells directly and through distributors. Call headquarters to be put in touch with a regional sales office.

Recognition Equipment
2701 East Grauwyler
Irving, TX 75061
(214) 579-6000

Tartan Plus is a proprietary network used primarily for data entry and data capture applications. It accommodates Z80-A 8-bit microprocessors. It will support 3,300 Z80-A microprocessors in one location. A user can attach other devices from Recognition Equipment to each processor such as a terminal, disk drive, tape drive, or an IBM Personal Computer. Tartan Plus was introduced in 1982. Recognition Equipment has installed 475 networks. It sells through a direct sales force in the United States.

Syntrex Inc.
246 Industrial Way West
Eatontown, NJ 07724
(201) 542-1500

SynNet attaches Syntrex's own systems, terminals, and file servers to a local area network. Over 800 workstations are supported for office applications.

Sytek Inc.
1225 Charleston Road
Mountain View, CA 94043
(415) 966-7300

Sytek, Inc., was founded in 1979 by five engineers and scientists from Ford Aerospace and Communications Corporation, to provide communications and data security consulting services to the federal government, universities and Fortune 500 companies.

Sytek made the transition from consulting to manufacturing with LocalNet 20—the first commercially available broadband LAN—in 1981.

The industry's first commercial encrypted LAN was introduced by the company in 1983. In 1984, an alliance between Sytek and IBM produced the IBM PC network. Within that year, Sytek installed the industry's first 5,000 connection network.

By 1986, the median size of Sytek's installations had grown to 500 connections. By 1987, the company's first 10,000-connection network was installed, the result of a clear continuing focus on the special needs of the largest computer-using organizations.

Multiple-protocol architecture had become a part of the LocalNet System by 1987, and in 1988, a token-bus backbone was added, as well as Ethernet 802.4 subnets, MAC-level bridge with backbone routing, and remote network management.

Today, the company is positioned to supply comprehensive LAN system solutions to large organizations with complex, heterogeneous computing environments. Sytek integrates, configures, installs, and supports LAN solutions, including cable system design and installation. Training, network management and maintenance are also part of an extraordinarily complete offering of products and services.

3-Com Corp.
3165 Kifer Road
Santa Clara, CA 95052
(408) 562-6400

3-Com designs, manufactures, and markets Ethernet-based LANs for 16-bit and 32-bit personal computers and workstations. Network servers have built-in Localtalk and Token Ring products. 3-Com also has gateways that connect its personal-computer LANs to mainframes and minicomputers. 3-Com says, "There are 450,000 personal computers connected to more than 36,000 3-Com networks." 3-Com products are available through direct sales, resellers, dealers and value-added resellers, as well as national and international distributors. Dealers include Businessland, Computerland, Entre Computer Stores and Tandy Radio Shack Computer Centers.

Ungermann-Bass Inc.
3900 Freedom Circle
Santa Clara, CA 95054
(408) 496-0111

Since its founding in 1979, Ungermann-Bass, Inc. has been committed to developing general purpose, standards-based networking systems capable of interconnecting diverse types of information processing equipment from multiple vendors.

Its initial systems, Net/One, are capable of interconnecting mainframes, minicomputers, workstations, PCs, terminals, modems, and other devices.

Net/One supports the IEEE 803.5 Ethernet and 802.5 Token Ring networking standards. Ungermann-Bass began shipping Net/One in 1980; there are now more than 2,200 installations.

Another product line, MAP/One, is a standards-based data communications system for factory environments. MAP/One is based on the IEEE 802.4 protocols, called the Manufacturing Automation Protocol (MAP). Ungermann-Bass supplied the networking system for General Motors' GMT 400 Project, the world's first plant-wide implementation of MAP.

In January 1988, Ungermann-Bass announced the next generation of networking architecture with the introduction of Access/One. Serving as a platform for the delivery of network services, Access/One provides connectivity for asynchronous devices, 3270 terminals, PCs and other distributed devices via Ethernet and Token Ring over common, twisted-pair telephone wire.

Ungermann-Bass markets its products through a direct sales force, OEMs, and distributors. Contact the company or one of its local sales offices.

Wang Laboratories Inc.
One Industrial Way
Lowell, MA 01851
(617) 459-5000

The WangNet local area network is a broadband radio-frequency hardware medium for the concurrent exchange of data, text, graphics, electronic mail, and imaging. WangNet can meet communications requirements within a single- or multi-level building. An expanded design can serve a multiple-building site such as a college campus, hospital, military base, or large corporation. WangNet is designed to transport information across Wang and non-Wang equipment. It supports any device that is compatible with the IEEE 802.3 Ethernet standard. IBM's PC Network for IBM PCs, XTs, ATs and compatibles can be implemented as one of the services on the WangNet. Wang also offers a turnkey modular broadband network (FastLAN). FastLAN's architecture allows the user to implement WangNet at the user's own pace: one department, one floor, or one building at a time.

Xerox Corp.
Xerox Center
101 Continental Boulevard
El Segundo, CA 90245-4899
(213) 333-7000

Xerox's Ethernet is a local area network recognized as a world standard. It allows mixed office systems such as workstations, personal computers, word processors, laser printers, and scanners to read both text and graphics and to exchange information with each other. The network can link more than 40 Xerox products, plus equipment from other vendors, such as the IBM PC and DEC minicomputers. Ethernet's transmission speed is 10 Mbps. An Ethernet network can run through offices on a single floor, a whole building, or a few closely grouped buildings. A single Ethernet coaxial cable is 1½ miles long and can accommodate up to 1,024 workstations and other devices. One network can be attached to another, so that the number of workstations installed in any LAN is virtually unlimited. Xerox also provides a means for other kinds of networks to operate with Ethernet. These include PBX-switched phone lines, DEC NET, X.25 communications and IBM's SNA. More than 35,000 Ethernet local area networks have been installed by Xerox and other Ethernet vendors. In addition, over 350 other manufacturers are now producing and installing Ethernet products around the world.

XYPLEX, Inc.
100 Domino Drive
Concord, MA 01742
(617) 371-1400
(800) 338-5316

The Xyplex communications server family allows RS-232 terminals, PCs, modems, and other asynchronous devices to connect to DEC VAX/VMS hosts, and to any host computer running TCP/IP Telnet protocols.

Xyplex's MAXserver 5000 is the company's latest Ethernet product, and it serves up to 600 users on a single 60-inch rack. It features the smallest size of any server on the market, and can be custom configured simply by plugging in user-installable and replaceable module cards. Its open architecture is ideal for growing networks, allowing you to add options and users as need requires.

The MAXserver is backed by a three-year parts and labor warranty. Mean time to repair for the MAXserver ranges from three minutes to five minutes maximum, so downtime is dramatically reduced.

Appendix II

Third-Party Vendors of Local Area Networks

Bell Atlantic Network Services Inc.
1310 North Courthouse Road
Arlington, VA 22201
(703) 974-3000

The Central Office-Based Local Area Network (C.O. LAN), introduced in April 1985, adds new data capabilities to Centrex. C.O. LAN eliminates the need for separate, expensive cabling by utilizing existing Centrex wiring for voice and data communications. An inexpensive data / voice multiplexer at the user's location is matched with a similar device in Bell Atlantic's Central Office. When a user sends data from a personal computer or other communications device, it goes through the Central Office, which routes the signal back to the addressed destination. C.O. LAN allows business users to connect personal computers, hosts, and peripherals, using normal telephone wires. It is an asynchronous-to-asynchronous communications method. Any standard speed from 300 bps to 19.2 Kbps is supported. It is intended for campus sites or for buildings across towns or between towns. C.O. LAN is marketed and installed by the local telephone companies under Bell Atlantic, including Bell of Pennsylvania, C&P Telephone, Diamond State Telephone, and New Jersey Bell. For further information, contact your local telephone company.

Businessland
101 Ridder Park Drive
San Jose, CA 95131
(408) 437-0400

Your local Businessland retail store is a supplier of personal-computer local area networks. Businessland installs and supports network products from Novell, IBM, and 3-Com. It will also connect Macintosh personal computers. Businessland has been offering networking since 1982 and has more than 4,000 personal-computer

LAN installations nationwide. Businessland has 93 branches across the U.S. as well as 9 locations in the United Kingdom.

GE Calma Company
501 Sycamore Drive
Milpitas, CA 95035
(408) 434-4000
(800) GE CALMA

GE Calma Company, a wholly owned subsidiary of General Electric, is a third-party vendor of local area networks. It will install and support LANs for small to very large Fortune 500 companies, with applications geared to computer-aided design and computer-aided manufacturing (CAD/CAM). The end-user must purchase GE Calma Company's CAD/CAM software and then GE Calma will negotiate the OEM agreements with key LAN vendors. The network hardware will be transparent to the end-user. Calma's software runs on the DEC VAX systems, Apollo and Sun microsystems. For further information, contact GE Calma Company.

ComputerLand Corporation
2901 Peralta Oaks Court
Oakland, CA 94605
(415) 487-5000

ComputerLand provides sales, service, and installation of personal computer-based LANs from several manufacturers, including IBM, Apple, Novel, 3-Com, AT&T, Classic Technology Corp., and AST Research.

ComputerLand also provides education, training, and after-sale support. With 800 stores in 33 countries, ComputerLand is the world's largest specialty retailer for personal computers. ComputerLand stores are franchised, and the LAN product lines that independent owners offer may vary from store to store. For more information, contact your local ComputerLand or call (415) 487-5000.

Contel Customer Support
245 Perimeter Center Parkway
Atlanta, GA 30346
(404) 395-8705

Contel performs all of the services needed to provide complete LAN solutions including analysis, planning, design and engineering, implementation, maintenance, and operation. The Contel product line covers the complete spectrum from fiber backbones for a campus to building distribution systems, to workgroup LANs. The products interconnect LANs, connect terminals and PCs to hosts, and provide PC/file server networking. Contel is not a manufacturer, but an integrator of products from 3-Com, Bridge Communications, Proteon, David Systems, SynOptics, and others. These products provide Ethernet, broadband, and 4-, 10-, and 80-Mbps token ring LANs on Ethernet cable, unshielded twisted pair, and fiber. For further information, contact the LAN department at Contel Customer Support division.

DMW Group
2020 Hogback Road
Ann Arbor, MI 48104
(313) 971-5234

DMW supplies network designs for large Fortune 500 clients and universities and multi-vendor sites. It assists with vendor bid proposals and helps implement the network.

Electronic Data Systems (EDS)
7171 Forest Lane
Dallas, TX 75230
(214) 661-6000

EDS is a single-call supplier of local area networks targeted toward the largest companies with multi-vendor environments. EDS will analyze and define the end user's needs and design, install, support, and maintain the network according to those requirements. The network hardware, for which EDS negotiates from different LAN manufacturers, will be transparent to the end user. For further information, contact EDS. EDS says it is the largest systems integration company and has been installing networks since 1962.

MicroAge Computer Stores Inc.
2308 South 55th Street
Tempe, AZ 85282
(602) 968-3168

MicroAge Computer Stores has 180 stores worldwide that offer full installation and support of personal computer-based LANs. Novel networks are its major offering; 3-Com products are also sold. MicroAge Computer Stores is a franchised network—the individual stores can choose their own product mix. For more information, call your local MicroAge store or the company's headquarters.

Network Solutions Inc.
8229 Boone Boulevard
Vienna, VA 22180
(703) 442-0400

Network Solutions, a full-scale integrator of LANs, designs and implements medium- to large-scale projects (networks that cost $100,000 or more). Network Solutions assesses a user's environment and selects the best LAN hardware to do the job. Much of Network Solutions' work has been in designing LANs for the federal government. It has a particular specialty in designing TCP-IP-based networks, but its expertise is not limited to that. It designs broadband, baseband, and fiberoptic networks and has used a diverse range of equipment manufacturers, including Ungermann-Bass, Bridge Communications, Digital Equipment, and 3-Com. For further information, contact the vice president for program development.

Nynex Business Information Systems Company
4 West Red Oak Lane
White Plains, NY 10604
(914) 993-3800

Nynex Business Information Systems Company offers a personal-computer LAN, based on products from 3-Com. Nynex Business Centers will design, install, maintain, and certify the network.

PacTel InfoSystems
1777 Botelho Drive
Walnut Creek, CA 94596
(415) 947-5000
(800) PACTEL-5

PacTel InfoSystems, a third-party integrator, offers a full range of LANs for a wide variety of purposes and uses a variety of LAN manufacturers' products. Network engineers will assess the end user's particular applications and propose solutions. End user applications can vary from small desktop publishing departments to campus-sized multi-vendor environments. Its personal-computer LANs include Appletalk for the Macintosh, IBM Token Ring, and 3-Com products using twisted-pair wire or coaxial cable. PacTel carries products from the following companies: Tops, Kinetics, DCA, Novell, Telebit and Farallon. PacTel also offers backbone LANs to tie multiple LANs together, using fiberoptic technology. The company has installed several hundred LANs, ranging from a three-node network up to a $1.2 million fiberoptic network. For more information, call the 800 number.

The Personal Computing Group
330 Seventh Avenue, 18th Floor
New York, NY 10001
(212) 714-1814

The Personal Computing Group has been installing 3-Com based personal-computer networks since 1983. The group will obtain the network equipment for the end user, install and maintain the network, and manage it. Thirty networks have been installed. For further information contact The Personal Computing Group.

Southwestern Bell
One Bell Center
Saint Louis, MO 63101

Southwestern Bell Telephone Company is offering a central-office-based LAN that integrates voice and data to Plexar (or Centrex) customers. This allows users to connect personal computers to terminals, hosts to hosts, terminals to hosts, and stations to stations. User-transmitted information is routed to its destination via the central office switch of the local telephone company. This approach is well suited to campus environments and users who have multiple locations served by the same telephone company. Southwestern Bell serves five states: Missouri, Kansas, Oklahoma, Arkansas, and Texas. For further information, contact your Southwestern Bell Telephone Company account executive.

Tele-Engineering
2 Central Street
Framingham, MA 01701
(617) 877-6494
(800) TEC TELE
FAX (617) 788-0324

An engineering and turnkey operation that designs and installs large broadband high-capacity LANs or fiberoptic LANs, Tele-Engineering offers computer connectivity and concentrates on large installations such as college campuses, large manufacturing plants, hospitals, and military installations. Tele-Engineering has installed 80 LANs. It builds a transmission system transparent to the end-user. For further information, contact the company directly.

Sources For Additional Information

Books

Guide to Local Area Networks by T.J. Byers. Englewood Cliffs, NJ: Prentice-Hall, 1984.

Local Area Networks: An Introduction to the Technology by John E. McNamara. Burlington, MA: Digital Press, 1985.

Local Area Networks: A Study and Analysis. New York: Cegmark International, Inc., 1985.

Local Area Networks: A User's Guide for Business Professionals by James Harry Green. Glenview, IL: Scott, Foresman and Company, 1985.

Local Area Networks: Selection Guidelines by James S. Fritz, Charles F. Kaldenbach and Louis M. Progar. Englewood Cliffs, NJ: Prentice-Hall, 1985.

The Practical Guide to Local Area Networks by Rowland Archer. Berkeley, CA: Osborne McGraw-Hill, 1986.

Magazines

Computer and Communications Decisions, Hasbrouk Heights, NJ.

ComputerWorld, Manhasset, NY.

Data Communications, New York.

Datamation, New York.

MIS Week, New York.

Index